PROBABILITY IN APL

Other APL PRESS books

Algebra—An Algorithmic Treatment
APL and Insight
APL in Exposition
APL Language
APL Quote-Quad: The Early Years
Calculus in a New Key
Elementary Analysis
The Four Cube Problem
Introducing APL to Teachers
Introduction to APL
Introduction to APL for Scientists and Engineers
Resistive Circuit Theory
A Source Book in APL
Starmap
1980 APL Users Meeting
1982 APL Users Meeting
Volume I - Applications Sessions
Volume II - Special Technical Notes

PROBABILITY IN APL

Linda Alvord

Illustrations by **Dick Davies**

APL PRESS   Palo Alto

Library of Congress Cataloging in Publication Data

Alvord, Linda.
   Probability in APL.

   Includes index.
   1. Probabilities—Data processing. 2. APL (Computer program language) I. Title. II. Title: Probability in A.P.L.
QA273.A48   1984   519.2'028'5424   84-16902
ISBN 0-917326-16-4

ISBN: 0-917326-16-4
APL PRESS, 220 California Ave., Suite 201, Palo Alto, CA 94306-1683
©1984 by APL PRESS.
All rights reserved.
Printed in the United States of America

PREFACE

Thanks are in order to the many members of the growing APL community who have assisted me to develop educational materials during the past sixteen years. Ken Iverson has been my model of a teacher who along with Al Rose, Adin Falkoff, Len Gilman and Paul Berry have been most helpful from the beginning.

My colleagues in education including—but not limited to—Lola Bradway, Marge McElroy, Elizabeth Ann Mulford, Mary Rossi, Bruce Meserve, Max Sobel, Dorothy Roberts, Perry Tyson, Lynette Birkins, Charlie Waters, Tama Traberman, Jon Klimo, George Pallrand, Carolyn Maher, Claire Jacobs and Joseph Rosenstein gave me much support in these efforts.

New friends in APL including Keith Smillie, Dave Saunders, Alex Morrow, Carl Cheney, Arthur Whitney, Howard Peelle, Jim Swift, Henri Schueler, Marguerite Boisvert and John McPherson have taught me many ways to present ideas more clearly and concisely. Alex helped this development of probability greatly with the delightful *PERMALT* function. Howard Peelle and Henri Schueler provided ideas for the *PERM* function. The literally hundreds of students and staff with whom I work each day enrich my understanding and motivate me to find better ways to teach.

My family wonders if I have ceased to exist since they haven't seen me in over a year. Arlene Azzarello and Gene McDonnell know I do exist because of the many times they have assisted me as they have skillfully introduced me to the process of text preparation. Their words and ideas are found throughout these pages, but I take full responsibility for any errors that exist.

Linda Alvord
Glen Gardner, NJ 08826
May 1984

CONTENTS

1. ARRAYS /11
2. THE ROLL FUNCTION /15
3. THE RESHAPE FUNCTION /19
4. CREATING CHARACTER ARRAYS /23
5. POSSIBLE OUTCOMES /29
6. POSSIBLE OUTCOMES AS CHARACTERS /39
7. ALL POSSIBLE COMBINATIONS /49
8. PROBABILITY /55
9. BINOMIAL DISTRIBUTIONS /65
10. COMBINATIONS /75
11. PERMUTATIONS /81
12. GEOMETRIC DISTRIBUTIONS /95
13. THE DEAL FUNCTION /103
14. THE WORLD SERIES PROBLEM /111
15. THE FIRST ACE /117
16. THE BIRTHDAY PROBLEM /123

APPENDIX A: DEFINING FUNCTIONS /129
APPENDIX B: DEFINED FUNCTIONS /133
APPENDIX C: ANSWERS /137
APPENDIX D: APL INDEX /141

1/ ARRAYS

$$\alpha \div \omega$$
$$\div \omega$$

What chance do you have of winning a prize in a raffle when 32 tickets have been sold, the prizes are to be given to the first 7 tickets picked, and you hold 1 ticket?

The part of mathematics dealing with questions like this is called *probability*. The mathematical word for "chance" is, in fact, "probability," and probability is the mathematical study of the laws of chance. This booklet uses APL (A Programming Language) to teach the mathematical aspects of the laws of chance. There are several advantages of this approach. Many of the traditional notions of probability theory have no consistent symbolic language for their expression. However, these concepts can be written in the executable notation, APL. As with any language, APL is probably best learned by use in a real environment where its full meaning becomes apparent empirically. The showcase of probability theory is an excellent one for exhibiting the power and consistency of APL. This functional language organizes and clarifies the theory in return.

Since APL is a language, suppose we consider how most children begin to enjoy thinking, speaking, reading and writing using English. Usually, words are repeated aloud for the young child. Complexity grows as "cookie" is replaced by "Bobby wants a cookie." Sentence construction becomes known through use. In a similar fashion, children are introduced to books with a small selection of words and a simple syntax for the arrangement of the words as elements of sentences. These books are read and reread with growing delight. We will proceed in a fashion that parallels this natural way of learning a language.

In this chapter, we develop some essential concepts about arrays. This background provides a framework for using data in the study of probability. Comparisons with familiar notions in geometry can make it easier to create a mental image of arrays of data.

Geometry is a study of elements in space. The familiar vocabulary for the subsets of space includes words like "point," "line," "plane," "3-space" . . . "n-space." The dimensions in space associated with these words are 0, 1, 2, 3 . . . n, respectively. In APL, the language of *arrays* provides counterparts to the language of space. Arrays are ordered collections of data elements. Arrays can be scalars, vectors, matrices, 3-axis arrays . . . n-axis arrays. The number of *axes*, or dimensions, of the corresponding arrays are 0, 1, 2, 3 . . . n.

A *scalar*, a single element of data roughly corresponding to a dimensionless point in geometry, is represented by a single coin in Figure 1. A *vector*, or 1-axis array, is shown symbolically as several links in a chain. Each of the links represents a data element, so the vector in Figure 1 is a 5-element vector. A *matrix*, or 2-axis array, corresponds to a plane in geometry. The matrix in Figure 1 has 6 elements arranged in 3 rows and 2 columns. The 3-axis model in Figure 1 has 24 elements arranged in 2 planes of 4 rows and 3 columns each.

This somewhat meager vocabulary like "mama," "dada," "boat," "doggie," "me" and "cookie" gives us the beginnings of language. Next, we need a working environment for the development of an executable notation for the language. Just as we might begin an essay or letter by taking a blank piece of paper, we use a *system command* to obtain a clear workspace on an APL computer:

```
    )CLEAR
CLEAR WS
```

All APL system commands begin with a *right parenthesis*) and have an effect on the working environment.

In the example, we entered the system command $)CLEAR$. The APL computer supplied the message $CLEAR\ WS$ at the left margin of the display, indicating that it had supplied the equivalent of a fresh sheet of paper.

Essentially, APL is an executable notation in which *functions* that apply to arrays of data can be used to produce new arrays of data. The arrays that the functions are applied to are called *arguments*. A function is *monadic* or *dyadic* depending upon whether it requires one or two array arguments. Suppose we have a pocket calculator which has a reciprocal key. For that calculator, *reciprocal* is a *primitive* function. All of APL's primitive functions are available in a clear workspace; that is, they don't have to be created. They are most often represented by a single symbol. Most APL function symbols can be applied both to one argument (monadically) or to two arguments (dyadically). For example, the APL symbol ÷ for the dyadic function *divided by* applies to both a left and right argument as in conventional mathematics:

```
    3÷5
0.6

    1÷5
0.2
```

12/ Arrays

Dimensions: 0
Name in space: Point
APL array name: Scalar

Dimensions: 1
Name in space: Line
APL array name: Vector

Dimensions: 2
Name in space: Plane
APL array name: Matrix

Dimensions: 3
Name in space: 3-space
APL array name: 3-axis array

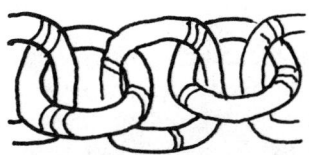

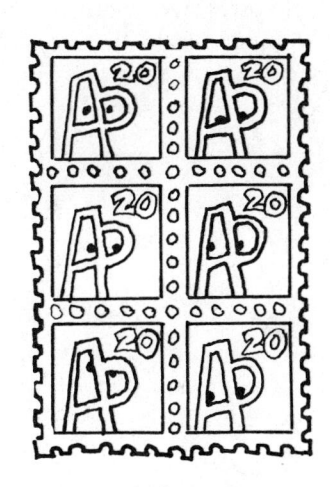

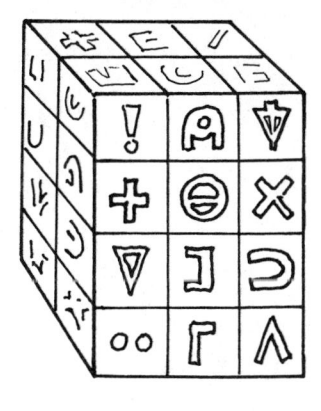

Fig. 1 - Shapes of Arrays

When we want to indicate the left and right arguments that a function applies to generally, rather than specifically, we will denote them by α *alpha* and ω *omega* respectively. Thus, we would represent the APL division function by α÷ω. The monadic use of an APL function requires only a right array argument:

```
      ÷5
0.2
```

The monadic *reciprocal* function ÷ω is a subset of division in which the implied value of α is 1.

In the preceding three examples, scalar arguments produce scalar results. Functions can be extended to apply to all arrays. In the next two chapters we create vector, matrix and 3-axis arrays. Then, we apply functions to them to investigate the laws of chance.

2/ THE ROLL FUNCTION

$?\omega$

In probability theory, we explore events which are affected by chance. We can consider the tossing of coins or rolling of dice. We might twirl a spinner, examine a lottery drawing or analyze the behavior of a "one-armed bandit." APL functions allow us to simulate such experiments.

Suppose we are interested in simulating the toss of a coin. We can use the computer to generate random integers with the APL *roll* function, $?\omega$. The right argument ω must be an array of positive integers. We can use the integer 2 to represent the two possible outcomes:

```
      ?2
2

      ?2
1
```

We can choose to consider a result of 1 to stand for a tail, and the other possible result, 2, to represent a head. Successive entries of $?2$ will yield either a 1 or a 2, in a random fashion.

A typical die is a cube which has six numbered faces. A roll of 6 simulates a roll of the die:

```
      ?6
3

      ?6
2
```

Each time we enter $?6$, the APL computer will generate at random one of the first six positive integers. Successive experiments with the roll function provide results similar to those we would get if we were actually rolling a die.

To simulate the selection of a winner of a door prize by drawing a ticket from a bowl with 200 tickets numbered from 1 to 200, we can have the computer generate one of the first 200 positive integers:

```
      ?200
183
```

The Roll Function /15

When considering situations involving the random selection of equally likely events, we can simply assign them counting numbers. If we want to simulate choosing a dessert from a menu of four equally desirable choices of pie, cake, ice cream or rice pudding, we can assign a counting number to each of the choices in order. Then we use a roll:

```
      ?4
3
```

and use the result to indicate the choice of the third dessert, or ice cream.

We can simulate several events simultaneously by using a vector argument. Rolling a vector of five 2s simulates the tossing of one penny five times:

```
      ?2 2 2 2 2
1 1 2 2 1
```

We can interpret the 5-element result in the example as a tail, tail, head, head and tail.

(Notice that we included blanks between the 5 elements of the vector argument when we entered it. Without the blanks, the argument is a single scalar:

```
      ?22222
14687
```

In this example, the result is randomly selected from the first 22,222 positive integers.)

A computer simulation of rolling a die ten times is:

```
      ?6 6 6 6 6 6 6 6 6 6
3 5 4 1 2 6 5 3 3 3
```

Looked at another way, this APL expression could also represent rolling ten dice once.

A three-digit numeral for a lottery might be chosen by putting nine Ping-Pong balls numbered from 1 to 9 into each of three containers. We could then select one Ping-Pong ball from each container. The computer could simulate the drawing process:

```
      ?9 9 9
4 2 9
```

16/ The Roll Function

Sometimes, it is useful to simulate mixtures of random events. Rolling two dice and selecting three Ping-Pong balls from the three containers are simultaneously simulated by a roll of two 6s and three 9s:

```
      ?6 6 9 9 9
5 2 9 3 4
```

EXERCISES

The exercises for each chapter are designed to illustrate the concepts presented in it. They might also suggest additional simulations useful in solving other types of problems. Answers are in Appendix C.

2.1 What APL expression would simulate a spinner which might come to rest in any one of five equal regions?

2.2 Simulate rolling an octahedral die.

2.3 Roll the octahedron a dozen times.

2.4 Roll a tetrahedron five times.

2.5 Roll two tetrahedrons and two octahedrons once.

2.6 Roll one tetrahedron and one octahedron twice.

2.7 Have the computer decide whether to swim, bike, play tennis, weed the garden or read a good book on a nice summer day. Consider all choices equally likely.

2.8 A classic problem in probability theory involves a random walker. Consider this walker in a large city with square blocks that are formed by streets that run North, South, East and West. If a random decision is made at each intersection, simulate a ten-block walk. It might be interesting to plot several walks on 1/4-inch graph paper.

3/ THE RESHAPE FUNCTION

⍺⍴⍵

"But what about rolling a *pair* of dice ten times?" Mathematicians typically ask such questions because it is often helpful to extend a problem in order to discover generalized solutions. We've already seen that APL functions take array arguments and provide array results. Thus, a matrix with two rows and ten columns of data suggests a strategy for rolling two dice ten times.

The APL *reshape* function ⍺⍴⍵ creates data arrays. We specify the shape of the array in the left argument ⍺. A 2-element vector ⍺ determines the length of each of the two axes in the matrix result of dyadic ⍴. For example, when ⍺ is the vector 2 10, the resulting matrix will consist of two rows with ten columns each. Twenty data elements are required to fill the twenty positions in this matrix. The right argument of ⍴ supplies these elements.

Since each die has six sides, a matrix having two rows (for two dice) of ten 6s (for ten rolls) is an appropriate argument for the roll function. We create this matrix as follows:

```
      2 10⍴6
6 6 6 6 6 6 6 6 6 6
6 6 6 6 6 6 6 6 6 6
```

The reshape function extends the scalar 6 in ⍵ to each position in the resulting matrix. That is, it is not necessary to enter twenty 6s in order to generate a 2-by-10 matrix of 6s.

Using the matrix result above as ⍵ for the roll function simulates rolling the pair of dice ten times:

```
      ?2 10⍴6
3 1 2 4 3 1 5 2 4 6
1 3 4 2 2 5 1 3 3 4
```

Note that the roll function applies to *the entire expression to the right of it*. This is true of all APL functions. Thus, the result of 2 10⍴6 is the right argument of ? . The roll function applies to each element in the 2-by-10 matrix of 6s created by the reshape function.

If we imagine a simulation with ten rows and two columns, we can interchange the lengths of the two axes:

```
      ?10 2ρ6
3 1
1 2
2 1
4 5
2 6
6 6
5 5
3 5
2 4
1 3
```

Remember, each use of the roll function generates random results.

Suppose you actually roll three dice four times. You can use the reshape function to create a matrix of your empirical data. Specify the shape of the matrix in α, and enter all 12 data elements in ω. For example:

```
      3 4ρ1 3 3 4 3 1 4 1 6 2 1 6
1 3 3 4
3 1 4 1
6 2 1 6
```

A 5-row and 3-column matrix requires 15 elements. If you supply fewer than 15 data elements in ω, the ones you have entered are used cyclically:

```
      5 3ρ6 6 2 7
6 6 2
7 6 6
2 7 6
6 2 7
6 6 2
```

```
      5 3ρ6 6 2
6 6 2
6 6 2
6 6 2
6 6 2
6 6 2
```

We can use the preceding matrix to simulate five events each involving the roll of two dice and the toss of one coin. Once we create the

20/ The Reshape Function

appropriate array, the simulation follows easily:

```
      ?5 3ρ6 6 2
3 4 2
4 5 1
5 5 1
2 1 2
4 6 1
```

Earlier we talked about 3-axis arrays. Let us assume that three people roll two dice four times. The 3-element vector we select for α establishes the shape of the array and the lengths of its three axes:

```
      3 2 4ρ6
6 6 6 6
6 6 6 6

6 6 6 6
6 6 6 6

6 6 6 6
6 6 6 6
```

The computer displays a representation of the 3-axis result. The matrices (analogous to planes in geometry) are separated by blank lines. Each matrix has two rows and four columns. A roll of this array provides the desired result:

```
      ?3 2 4ρ6
3 1 4 2
2 5 1 3

3 1 4 4
5 4 1 2

6 3 2 1
6 1 3 4
```

The first matrix represents the first player's rolls of the dice. The simulations of the second player's rolls appear in the second matrix. Similarly, the third player's are shown in the third matrix.

Although it might seem obvious, it might be helpful to point out that the reshape function is also valuable in creating vectors with many like

The Reshape Function /21

elements. It allows us to enter data efficiently:

```
    15⍴6
6 6 6 6 6 6 6 6 6 6 6 6 6 6 6
```

The 15-element vector result can be used to simulate rolling 15 dice once:

```
    ?15⍴6
3 1 4 2 2 5 6 3 3 1 4 2 3 5 1
```

Using reshape, we can create any numerical array. These scalar, vector, matrix, 3-axis . . . n-axis arrays can all be used with the roll function to simulate experiments of chance. By actual experiments and simulations, you can gain an intuitive and empirical grasp of probability theory.

EXERCISES

3.1 Roll a tetrahedron.

3.2 Roll five tetrahedrons.

3.3 Roll 100 tetrahedrons.

3.4 Roll a pair of tetrahedrons ten times.

3.5 Simulate five players each rolling three tetrahedrons six times.

3.6 Six people enter a store and select a beverage of soda, coffee or tea as well as a dessert of pie, cake, ice cream or rice pudding. Simulate this using a matrix.

3.7 A new car is available with a white, blue, beige or red interior. The body colors are red, blue, rust, ivory or gold. Simulate ten possible color combinations for this car.

4/ CREATING CHARACTER ARRAYS

$$\alpha[\omega]$$
$$\alpha \leftarrow \omega$$

In APL, the elements of arrays are either numerical or character. The arrays we've seen so far were numeric. *Character arrays* are composed of symbols or characters, but are otherwise similar to numerical arrays. Character arrays can be scalars, vectors, matrices, 3-axis arrays ... *n*-axis arrays. You enter a character scalar by typing an APL quote ' followed by the character or symbol and another APL quote:

```
      '*'
*
```

The quotes aren't displayed in the result, but they are always implied in character arrays. The following 7-element character vector entered between quotes shows that a blank is a character:

```
      'HOW SO?'
HOW SO?
```

A character vector is a valid right argument for the reshape function. A numeric vector α and a character ω create a character matrix:

```
      6 7ρ'MY CAT'
MY CATM
Y CATMY
 CATMY
CATMY C
ATMY CA
TMY CAT
```

This method of generating character arrays extends to arrays of any shape.

Suppose we simulate tossing three pennies eight times, and we want to display the result using the characters *T* and *H* rather than the numbers 1 and 2. First, we produce a numerical array:

```
    ?3 8ρ2
1 2 2 1 1 2 2 1
2 1 1 2 1 2 2 1
1 1 2 1 1 2 1 2
```

Then we can use this numeric result as the right argument of the *indexing* function α[ω] to produce a character representation of the simulation. The numerical array of index values in ω selects elements according to their *position* in the character vector in α. The indexing function requires both a left [and a right bracket] to delimit the expression provided in ω.

The numeric result of ?3 8ρ2 shown in the preceding example would produce the following character representation when used as the right argument of the indexing function:

```
    'TH'[?3 8ρ2]
THHTTHHT
HTTHTHHT
TTHTTHTH
```

The *T* is selected wherever there is a 1 in the numerical array in ω. The second character in α, the *H*, is selected by each 2 in ω. Since the roll function randomly produces the array of index positions, entering the same expression again is likely to produce a different result. For example:

```
    'TH'[?3 8ρ2]
TTHHTHTH
HHTTTHTT
HTTHTHHT
```

The directions taken by the random walker who traveled ten city blocks can be simulated with the expression:

```
    ?10ρ4
4 1 3 2 2 1 3 2 4 4
```

We can generate a character array result of the ten-block random walk by indexing the character vector 'NSEW' with a numeric vector. This simulation expresses the sequence of directions as characters representing the four cardinal points:

24/ Creating Character Arrays

```
      'NSEW'[?10ρ4]
ESSNESSNEW
```

Of the APL functions we have discussed, only the roll function and the index function are based upon counting. Their behavior depends upon the value of a *system variable*. The names of APL system variables begin with the *quad* symbol ☐. System variables determine certain characteristics of a workspace. The system variable ☐*IO, index origin,* has a value of one when we create a clear workspace.

```
      )CLEAR
CLEAR WS

      ☐IO
1
```

When ☐*IO* is one, all APL functions which involve counting do so by starting with the number 1. We can choose to condition the workspace so that all counting begins with zero instead. To assign a new value to the index origin, we use the *specification* function α←ω. The value in ω is assigned to α:

```
      ☐IO←0
```

Now the roll function applied to 4 yields 0, 1, 2 or 3:

```
      ?10ρ4
3 0 1 2 1 3 0 0 1 2
```

But no difficulty occurs in indexing! The four elements in α are also counted using zero as the origin.

```
      'NSEW'[?10ρ4]
WWNESENSEW
```

When you create character array simulations, the index origin is not apparent.

Although it has no obvious impact on the character arrays we have developed, there is a significant reason for changing the index origin to zero. Consider the array of index values for tossing three pennies eight times in zero origin:

Creating Character Arrays /25

```
        ?3 8⍴2
1 1 0 0 0 1 0 0
0 1 0 0 1 0 1 1
0 1 1 0 1 1 1 1
```

The sum of the numbers in each column of the matrix is the number of heads obtained in each toss. In the earlier simulations in origin one, the sum of the values in each column was not so meaningful. Each toss can result in 0, 1 or 2 heads. In many probability experiments, zero is a plausible result. It indicates no "successful" events. In fact, counting from zero is the most efficient way to treat most concepts in probability. From now on we will use zero as our index origin unless noted otherwise.

Here are some more experiments you might find enjoyable. You can create some random "words" as follows:

```
        'ABCDEFGHIJKLMNOPQRSTUVWXYZ'[?10 4⍴26]
MNQT
VIEL
PRNQ
SURT
MMRA
TUPP
BNZV
YTLC
FJRE
IXKD
```

Or, you can test your luck by simulating five pulls on the lever of a one-armed bandit. Assume the slot machine has three dials with six different symbols on each dial:

```
        '*□△O∇T'[?5 3⍴6]
△∇T
O*O
∇□T
TOO
□*△
```

On a lighthearted note, here's a small "Greek salad." (We leave it to you to "toss" your own for two!)

26/ Creating Character Arrays

```
          'αωριε'[?2 4 4ρ5]
αωρρ
ερωε
ιεια
αερω

ωερι
ιεαε
ρρωε
ιαεω
```

EXERCISES

4.1 Toss 4 pennies 16 times. Express the result using the characters H and T.

4.2 Randomly select one of the five weekdays.

4.3 From the 13 clubs in a standard bridge deck, randomly select 5 cards. Assume that the selected card is replaced among the 13 cards before each new selection. This is called *with replacement* in probability theory.

4.4 Select, with replacement, 3 rows and 27 columns of letters from the vector `'COB'`.

4.5 The expression `'COMPUTER'[5 7 4 2 3 6 5]` has what result? *Hint:* Check your index origin if you don't get a familiar word.

4.6 Select, with replacement, four colors from the seven colors of the rainbow.

4.7 A large jar contains equal quantities of pennies, nickels, dimes and quarters. Randomly select five coins with replacement.

4.8 Create 25 random 5-letter words.

4.9 Simulate 20 pulls of the lever of a one-armed bandit with 8 possible symbols on its 3 spinning wheels.

Creating Character Arrays /27

4.10 Write one expression to simulate two players using the same "machine" in exercise 4.9. Have each player pull the lever ten times.

5/ POSSIBLE OUTCOMES

⍺T⍵
⍳⍵
f/⍵
PO:⍵T⍳x/⍵

We have discussed how to formulate an expression that simulates an experiment. Such an experiment tells us only what happened in one specific trial. We must take a different approach to better understand and anticipate what might happen in future trials.

The first step in making predictions is to explore the number of distinct possible outcomes. When one coin is tossed there are two possible outcomes. When two coins are tossed, it might appear that there are three possible outcomes: two heads, two tails or a head and a tail. Thus, we might expect two heads to occur one time out of three on the average. However, this is not the case.

The diagrams in Figure 2 illustrate that there really are a total of four possible distinct alternatives when two coins are tossed. That is, there are four paths from the results of the first toss to the results of the second. For example, a tail on the first coin and a head on the second has a different path through the diagram than a head on the first coin and a tail on the second. The other two possibilities are, of course, two heads or two tails. We shall assume in all these examples that we are using "fair" coins with which a head or a tail is equally likely. Thus, we would expect to throw two heads one time in four. This is different from our initial view of two heads in one out of three tosses.

The diagram for tossing three coins has an additional set of branches. Note that three of the eight paths in Figure 3 have two heads and one tail. "Tree diagrams" like those in Figures 2 and 3 show all the possibilities visually.

We can use the *encode* function ⍺T⍵ to express the information in a tree diagram. Encode generates the representation of the numbers specified by ⍵ in the base you indicate in ⍺.

```
      2 2 2T0 1 2 3 4 5 6 7
0 0 0 0 1 1 1 1
0 0 1 1 0 0 1 1
0 1 0 1 0 1 0 1
```

Flipping the
first coin

Flipping the
second coin

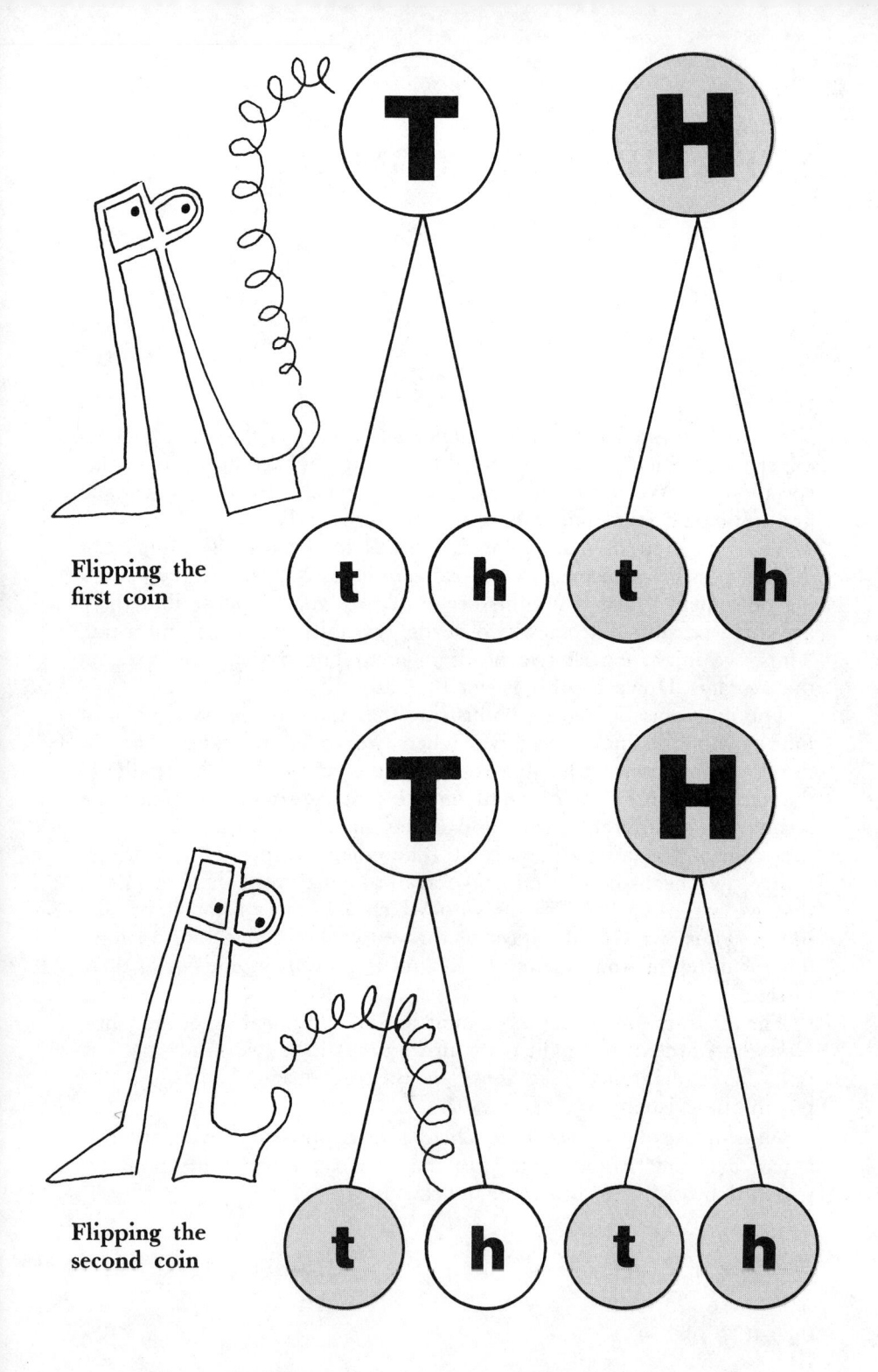

Fig. 2 - Possible Outcomes from Flipping Two Coins

Fig. 3 - Possible Outcomes from Flipping Three Coins

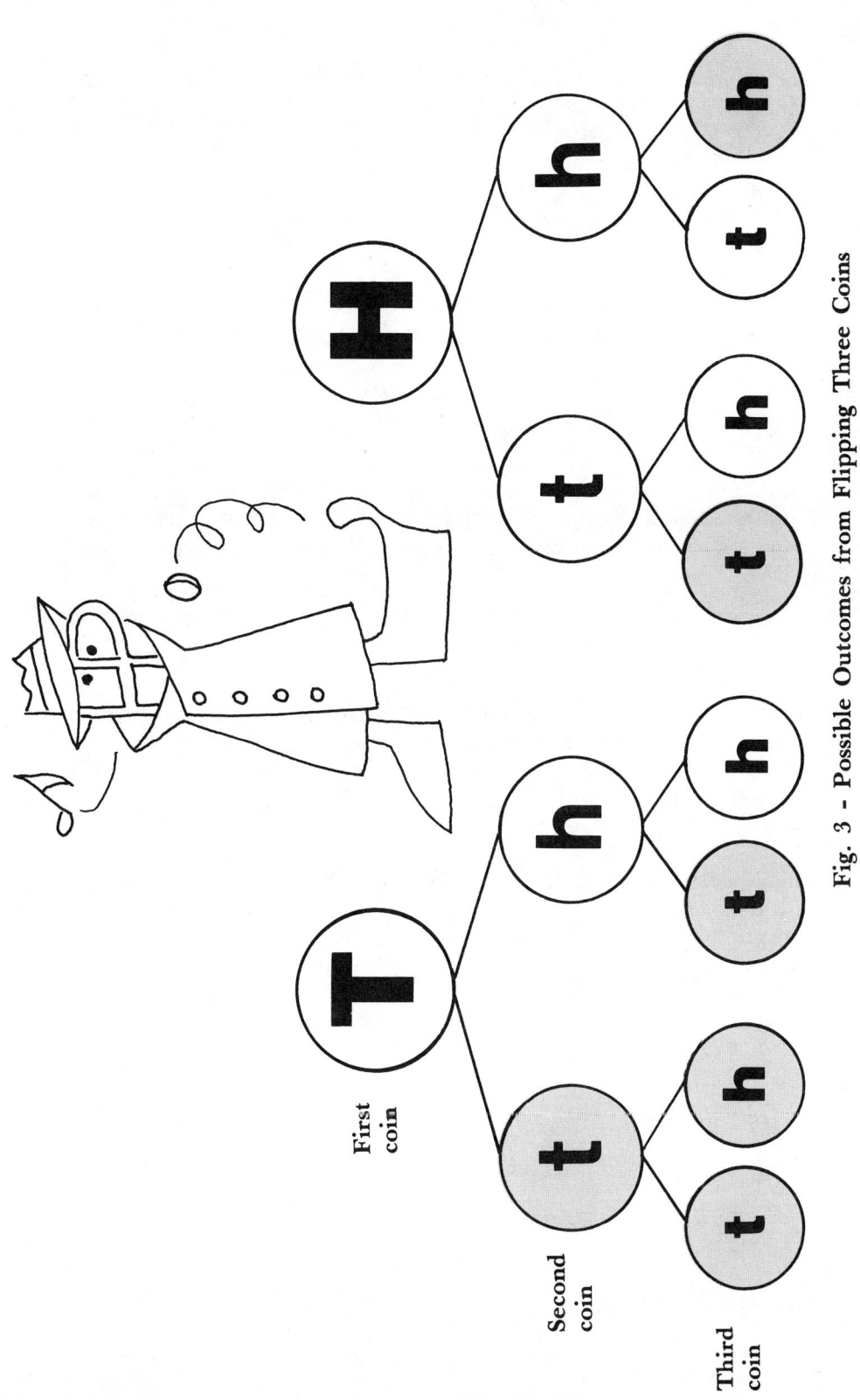

The columns of this matrix contain the information illustrated by the eight paths in Figure 3:

```
      'TH'[2 2 2T0 1 2 3 4 5 6 7]
TTTTHHHH
TTHHTTHH
THTHTHTH
```

Some additional examples of encode illustrate its characteristics. Suppose we want the digits of the number 375 to be represented in base 10:

```
      10 10 10T375
3 7 5
```

The left argument must contain as many 10s as there are digits in the representation of the number in base 10. Supplying fewer than three 10s in the left argument generates an incomplete base-10 representation of 375:

```
      10 10T375
7 5
```

However, if the leftmost element in α is zero, the result of the encode function will always supply a complete representation.

```
      0 10T375
37 5
```

This is a convenient rule to remember, especially when you're working with number systems less familiar than base 10. Suppose you want to generate the binary representation of the number 13. You could make the following series of tests to determine the number of 2s needed in α to completely represent 13 in base 2:

```
      0 2T13
6 1

      0 2 2T13
3 0 1

      0 2 2 2T13
1 1 0 1
```

32/ Possible Outcomes

```
      0 2 2 2 2T13
0 1 1 0 1
```

An easier rule to remember is that the product of the numbers in the left argument of encode must exceed the value of the number in the right argument. Thus, since 8 is less than 13, we know that three 2s will not give a complete binary representation of 13. However, 16 is greater than 13, so α must contain four 2s.

Notice how the encode function works when more than one number is in the right argument:

```
      10 10 10 10T2567 13 6589 379 8008
2 0 6 0 8
5 0 5 3 0
6 1 8 7 0
7 3 9 9 8
```

When five numbers are supplied in ω, the digits of the representation of each are displayed in the corresponding five *columns* of the matrix result.

The *index generator* function ιω creates a vector of the first ω integers starting with the index origin:

```
      ι16
0 1 2 3 4 5 6 7 8 9 10 11 12 13 14 15
```

To represent the first 16 counting numbers in base 2, we enter:

```
      2 2 2 2Tι16
0 0 0 0 0 0 0 0 1 1 1 1 1 1 1 1
0 0 0 0 1 1 1 1 0 0 0 0 1 1 1 1
0 0 1 1 0 0 1 1 0 0 1 1 0 0 1 1
0 1 0 1 0 1 0 1 0 1 0 1 0 1 0 1
```

This matrix result contains the counting numbers from 0 to 15 represented in the binary number system. The matrix also represents the 16 paths in a tree diagram for tossing 4 coins. Each element in the vector 2 2 2 2 indicates the number of choices for each successive set of branches in the tree diagram. The product of all the elements in 2 2 2 2 is 16, or the number of possible outcomes for tossing 4 coins.

We can obtain the total number of possible outcomes by using an *operator*. Operators modify the meanings of functions. The notation

f/ω represents the *reduction* operator. The reduction operator applies a dyadic function between each of the elements of a vector. To calculate total possible outcomes, we use the *times reduction* $\times/\omega$:

```
      ×/2 2 2 2
16
```

The advantage of this notation is that it "says," find the product of all the elements in the vector 2 2 2 2. A *plus reduction* $+/\omega$ of a vector of one-hundred 5s provides the sum of the elements in the 100-element vector:

```
      +/100ρ5
500
```

Plus reduction eliminates typing 99 plus signs!

The index generator function can be applied to the times reduction of 2 2 2 2:

```
      ι×/2 2 2 2
0 1 2 3 4 5 6 7 8 9 10 11 12 13 14 15
```

$\times/2$ 2 2 2 calculates the number of possible outcomes for tossing four coins. $\iota\times/2$ 2 2 2 generates the right argument of the encode function, giving the same result as 2 2 2 2⊤ι16:

```
      2 2 2 2⊤ι×/2 2 2 2
0 0 0 0 0 0 0 0 1 1 1 1 1 1 1 1
0 0 0 0 1 1 1 1 0 0 0 0 1 1 1 1
0 0 1 1 0 0 1 1 0 0 1 1 0 0 1 1
0 1 0 1 0 1 0 1 0 1 0 1 0 1 0 1
```

Notice that the entire matrix result has been created using the vector of possible outcomes 2 2 2 2 twice.

We can add our own *defined functions* to the set of APL primitive functions. Since the study of probability theory evolves from looking at the total number of possible outcomes of events governed by chance, we can save time by defining our own function PO. Thus, each time we need to generate a matrix of *possible outcomes,* we can execute PO with an appropriate argument and let the APL computer generate it for us.

$PO: \omega\top\iota\times/\omega$

A *colon* : separates the name of the defined function from the executable expression associated with it. The ω represents whatever right argument you specify when you use *PO* to generate a matrix of all possible outcomes. That is, ω stands for any vector of possibilities for consecutive events. Comments to assist you in defining your own functions are available in *APPENDIX A: Defining Functions*. Once you have defined a function, you can use it as you would APL primitive functions:

```
      'TH'[PO 2 2 2 2]
TTTTTTTTHHHHHHHH
TTTTHHHHTTTTHHHH
TTHHTTHHTTHHTTHH
THTHTHTHTHTHTHTH
```

Suppose you shuffle two decks of cards and place them face down next to each other. Then you turn over the top card of each deck. What are the possible outcomes for the suit of each of the two card sequences? There are four possible suits for each of the two cards, so the vector 4 4 describes the possibilities for the two consecutive events. We can use *PO* to generate the possible outcomes matrix. Then we can use the result of *PO* to index the character vector *'CDHS'* representing the four suits:

```
      'CDHS'[PO 4 4]
CCCCDDDDHHHHSSSS
CDHSCDHSCDHSCDHS
```

The computer executes the following sequence of steps. The defined function *PO* specifies the first three:

```
      ×/4 4
16
      ιx/4 4
0 1 2 3 4 5 6 7 8 9 10 11 12 13 14 15
      4 4⊤ιx/4 4
0 0 0 0 1 1 1 1 2 2 2 2 3 3 3 3
0 1 2 3 0 1 2 3 0 1 2 3 0 1 2 3
```

Indexing the character vector representing clubs, diamonds, hearts and spades provides the final result:

Possible Outcomes /35

```
      'CDHS'[4 4⍴⍳×/4 4]
CCCCDDDDHHHHSSSS
CDHSCDHSCDHSCDHS
```

Let's create a more interesting problem by allowing three choices for a first event, two for a second event and four for a third. To put the problem in an everyday context, we could be looking for all the possible outfits we could assemble from three different blouses, two different belts and four different skirts. Figure 4 illustrates the tree structure of the problem and shows the path the cartoon character took in selecting one of the 24 possible outfits. Drawing a tree diagram is a valuable aid until you gain facility in creating a matrix of possible outcomes.

The following sequence of steps illustrates the development of the corresponding matrix:

```
      ×/3 2 4
24

      ⍳×/3 2 4
0 1 2 3 4 5 6 7 8 9 10 11 12 13 14 15 16 17 18
      19 20 21 22 23


      3 2 4⍴⍳×/3 2 4
0 0 0 0 0 0 0 0 1 1 1 1 1 1 1 1 2 2 2 2 2 2 2 2
0 0 0 0 1 1 1 1 0 0 0 0 1 1 1 1 0 0 0 0 1 1 1 1
0 1 2 3 0 1 2 3 0 1 2 3 0 1 2 3 0 1 2 3 0 1 2 3
```

Displaying each partial result prior to using it as the argument of another function allows you to check your progress when you're developing a defined function. Once you've defined PO, you can enter its name followed by an argument. PO executes, in sequence, the three steps shown in the previous example:

```
      PO 3 2 4
0 0 0 0 0 0 0 0 1 1 1 1 1 1 1 1 2 2 2 2 2 2 2 2
0 0 0 0 1 1 1 1 0 0 0 0 1 1 1 1 0 0 0 0 1 1 1 1
0 1 2 3 0 1 2 3 0 1 2 3 0 1 2 3 0 1 2 3 0 1 2 3
```

The first row in the resulting matrix indicates which of the three different blouses is chosen. The second row applies to the two belts. The third row shows the skirt chosen. Each column represents a different outfit. All 24 different ensembles are provided in the result.

Fig. 4

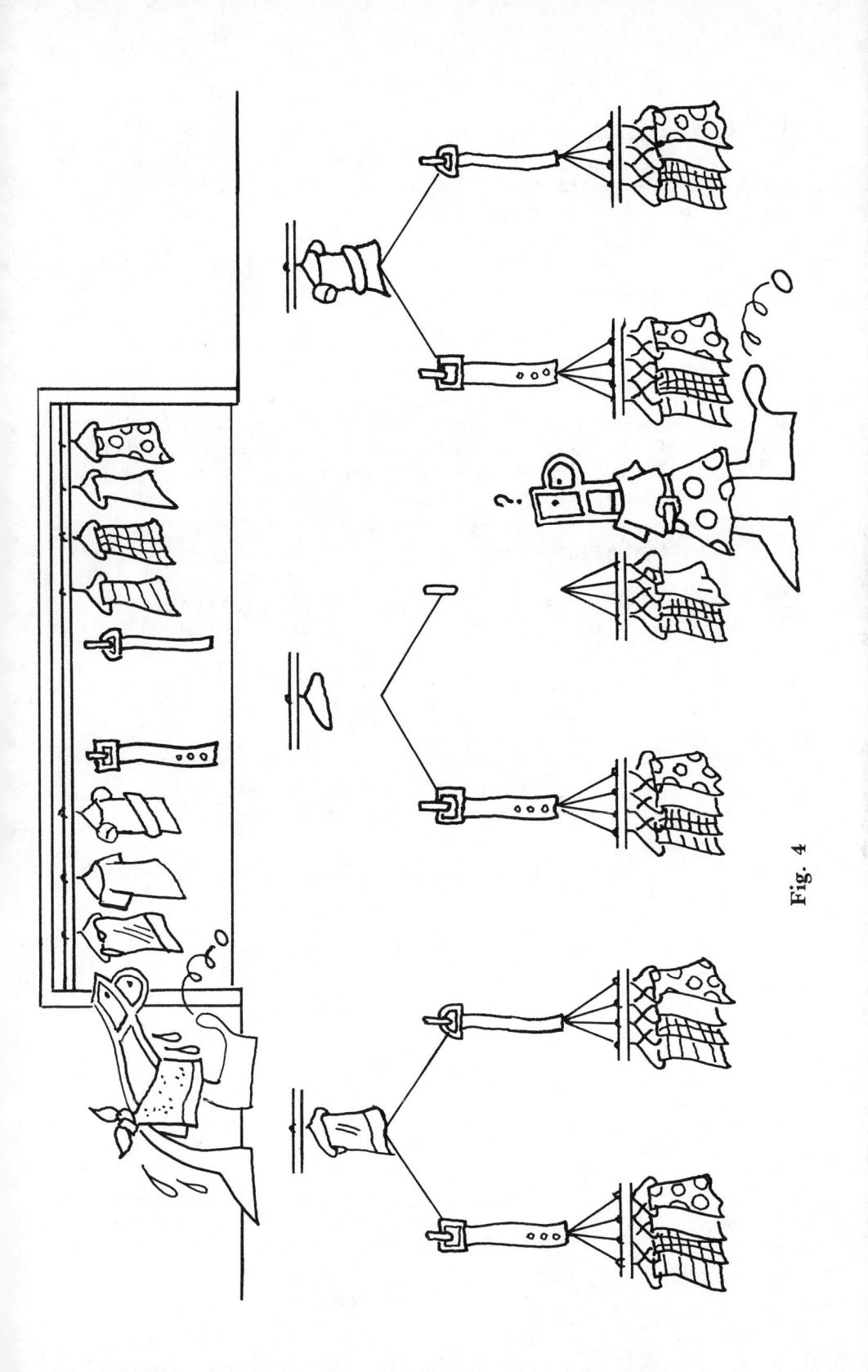

EXERCISES

5.1 Create the numerical matrix of 0s and 1s that shows the possible outcomes when 5 pennies are tossed.

5.2 Use the result from 5.1 to produce the corresponding character matrix of *T*s and *H*s.

5.3 Simulate tossing 5 pennies 32 times. Present your result as a character matrix of *T*s and *H*s.

5.4 In the local burger palace, it is possible to get a dinner of fish, chicken, ribs, burger or pizzaburger. Side orders of potato salad and cole slaw are available. The beverages are soda, tea and coffee. Create the matrix of possible outcomes for the vector 5 2 3 to represent all the ways a dinner, side order and a beverage can be selected.

5.5 What is the number of possible outcomes for a random walker who walked ten city blocks?

5.6 What is the number of possible four letter "words" randomly chosen from the alphabet? Repetition of letters is allowed.

5.7 What are all the possible outcomes if you select three flowers from a florist's stock of yellow roses, red carnations, white lilies, and purple hyacinths? Produce a character matrix of your result.

6/ POSSIBLE OUTCOMES

AS CHARACTERS

$CM: \lozenge(\phi\rho PO\omega)\rho^{-}1\downarrow+\backslash 0,\omega$
$POC: \alpha[(PO\omega)+CM\omega]$
α,ω
$f\backslash\omega$
$\alpha\downarrow\omega$
$\rho\omega$
$\phi\omega$
$\lozenge\omega$
$f.g$
$\circ.g$
$\alpha\circ.+\omega$
$CMALT: (^{-}1\downarrow+\backslash 0,\omega)\circ.+(\times/\omega)\rho 0$

As is often the case, solving one problem suggests another. We can produce character arrays of the possible outcomes of tossing coins by indexing a character vector '*TH*' with the matrix generated by *POω*. This is possible because each successive event is another toss of the coin. When the successive events are different, such as choosing blouses, belts and skirts to make an outfit, the result of *POω* must be modified before we can use it to produce a character array. Let's examine the matrix of possible outcomes again:

```
    PO 3 2 4
0 0 0 0 0 0 0 0 1 1 1 1 1 1 1 1 2 2 2 2 2 2 2 2
0 0 0 0 1 1 1 1 0 0 0 0 1 1 1 1 0 0 0 0 1 1 1 1
0 1 2 3 0 1 2 3 0 1 2 3 0 1 2 3 0 1 2 3 0 1 2 3
```

If the blouses are red, pink and yellow, we can choose the first letter of their respective colors to represent them. For the two different belts, we pick the symbols Δ and □. The skirts are gray, black, ivory and white. Thus, the vector '*RPYΔ□GBIW*' provides a symbol for each possibility. The index values in the first row of the matrix correspond to the positions of the characters representing the three different blouses in the vector '*RPYΔ□GBIW*'.

However, in the second row of the matrix we need 3s and 4s rather than 0s and 1s in order to index the symbols Δ and □ representing the two different belts. Thus, we must add a 3 to each element in the second row. In a similar fashion, we want elements of 5, 6, 7 and 8 instead of 0, 1, 2 and 3 in the third row in order to select the characters standing for each of the four skirts. Adding a 5 to each element in the last row would accomplish this.

Note that 3 is the first element in the right argument of PO and 5 is the sum of its first two elements. We can define a function which exploits this relationship of the values in the argument to PO and generate a *correction matrix* of the elements we want to add to the result of $PO\omega$:

$CM:\phi(\phi\rho PO\omega)\rho^{-}1\downarrow+\backslash 0,\omega$

```
      CM 3 2 4
0 0 0 0 0 0 0 0 0 0 0 0 0 0 0 0 0 0 0 0 0 0 0 0
3 3 3 3 3 3 3 3 3 3 3 3 3 3 3 3 3 3 3 3 3 3 3 3
5 5 5 5 5 5 5 5 5 5 5 5 5 5 5 5 5 5 5 5 5 5 5 5
```

Note that the size and shape of the correction matrix created by $CM\omega$ conforms to that of the matrix of index values given by $PO\omega$.

Before we analyze this algorithm, let's finish solving the problem. Adding the corresponding elements of these two matrices produces the following result:

```
      (PO 3 2 4)+CM 3 2 4
0 0 0 0 0 0 0 0 1 1 1 1 1 1 1 1 2 2 2 2 2 2 2 2
3 3 3 3 4 4 4 4 3 3 3 3 4 4 4 4 3 3 3 3 4 4 4 4
5 6 7 8 5 6 7 8 5 6 7 8 5 6 7 8 5 6 7 8 5 6 7 8
```

In APL, parentheses ( ) delimit a set of characters which are to be treated as a unit in evaluating an entire expression. First, we generate the result of CM 3 2 4. Then we add it to the result of PO 3 2 4.

A defined function $\alpha POC\omega$ permits us to represent possible outcomes directly as characters:

$POC:\alpha[(PO\omega)+CM\omega]$

```
       'RPY∆⎕GBIW' POC 3 2 4
RRRRRRRRPPPPPPPPYYYYYYYY
∧∧∧∧⎕⎕⎕⎕∧∧∧∧⎕⎕⎕⎕∆∆∆∆⎕⎕⎕⎕
GBIWGBIWGBIWGBIWGBIWGBIW
```

40/ **Possible Outcomes as Characters**

The defined function α*POC*ω is dyadic. Its right argument is a vector containing the number of possible outcomes for each event being considered. Its left argument is a character vector containing symbols corresponding to each possible outcome for each event in ω. The results of *PO*ω and *CM*ω are added to create the matrix of index values for the vector of symbols in α. The third column in the result of *POC* can be interpreted as a red blouse and an ivory skirt worn with the first of the two belts.

Now, let's examine the steps *CM* follows to generate the correction matrix.

CM:⌽(⌽⍴*PO*ω)⍴¯1↓+\0,ω

First, the scalar 0 is joined to the vector of possible outcomes 3 2 4 using the *catenate* function α,ω to produce a four-element vector:

```
      0,3 2 4
0 3 2 4
```

A monadic operator represented by the symbol \ is called *scan*. It is similar to the reduction operator and produces a vector of partial sums when used with the plus function:

```
      +\0,3 2 4
0 3 5 9
```

Remember, we need the first element, 3, and the sum, 5, of the first two elements from the right argument of *PO* to correct the index values in the second and third rows of the matrix created by *PO*. We catenated 0 to the elements in ω because we don't need to correct the index values in the first row.

Since the rightmost element in the vector result of plus scan is unnecessary, we can use the *drop* function α↓ω to eliminate it. The left argument of drop specifies the number of elements to be dropped from the vector in ω. If the number in α is positive, the elements are eliminated from the left end of the vector. If it is negative, they are dropped from the right end:

```
      ¯1↓+\0,3 2 4
0 3 5
```

The vector result of ¯1↓+\ω provides the numbers we need to add to each row of the matrix of index values generated by *PO*ω. Now we need to reshape the vector into a matrix of the same size and shape as that produced by *PO*.

We can use the APL *shape* function ρω to determine the number of rows and columns in the matrix created by *PO* regardless of the argument used with *PO*. That is, the shape function assesses the number of elements in each axis of the array in its argument:

```
      ρ0 3 5
3
```

Thus, the result of ¯1↓+\0,3 2 4 has one axis which contains three elements.

Let's look at the result of our possible outcomes function again:

```
      PO 3 2 4
0 0 0 0 0 0 0 0 1 1 1 1 1 1 1 1 2 2 2 2 2 2 2 2
0 0 0 0 1 1 1 1 0 0 0 0 1 1 1 1 0 0 0 0 1 1 1 1
0 1 2 3 0 1 2 3 0 1 2 3 0 1 2 3 0 1 2 3 0 1 2 3

      ρPO 3 2 4
3 24
```

The shape function reveals that the matrix produced by executing *PO* 3 2 4 has 2 axes with 3 elements in the first and 24 in the last; that is, 3 rows and 24 columns. So our goal is to build a 3-by-24 matrix containing the elements in the vector 0 3 5.

```
      3 24ρ0 3 5
0 3 5 0 3 5 0 3 5 0 3 5 0 3 5 0 3 5 0 3 5 0 3 5
0 3 5 0 3 5 0 3 5 0 3 5 0 3 5 0 3 5 0 3 5 0 3 5
0 3 5 0 3 5 0 3 5 0 3 5 0 3 5 0 3 5 0 3 5 0 3 5
```

As you can see, the reshape function provides a matrix of the proper dimensions, but the elements 0 3 5 are in the wrong order. We can use a combination of two APL functions to get all the 0s in the first row, the 3s in the second row and the 5s in the third. The *reverse* function ϕω reverses the sequence of elements along the last axis of the array in ω:

```
      ρPO 3 2 4
3 24
```

42/ **Possible Outcomes as Characters**

```
      ⌽3 24
24 3

      24 3⍴0 3 5
0 3 5
0 3 5
0 3 5
0 3 5
0 3 5
0 3 5
0 3 5
0 3 5
0 3 5
0 3 5
0 3 5
0 3 5
0 3 5
0 3 5
0 3 5
0 3 5
0 3 5
0 3 5
0 3 5
0 3 5
0 3 5
0 3 5
0 3 5
0 3 5
```

We're almost there! Note that each element of the vector 0 3 5 occupies a single column in the 24-by-3 matrix. We can now apply the *transpose* function ⍉ to interchange the rows and columns:

```
      ⍉24 3⍴0 3 5
0 0 0 0 0 0 0 0 0 0 0 0 0 0 0 0 0 0 0 0 0 0 0 0
3 3 3 3 3 3 3 3 3 3 3 3 3 3 3 3 3 3 3 3 3 3 3 3
5 5 5 5 5 5 5 5 5 5 5 5 5 5 5 5 5 5 5 5 5 5 5 5
```

So, here's the basis for our defined correction-matrix function:

Possible Outcomes as Characters

```
        ⍉(⌽⍴PO 3 2 4)⍴¯1↓+\0,3 2 4
0 0 0 0 0 0 0 0 0 0 0 0 0 0 0 0 0 0 0 0 0 0 0 0
3 3 3 3 3 3 3 3 3 3 3 3 3 3 3 3 3 3 3 3 3 3 3 3
5 5 5 5 5 5 5 5 5 5 5 5 5 5 5 5 5 5 5 5 5 5 5 5
```

We have been able to create the matrix in a sequence of steps that depend only on the vector of possible outcomes 3 2 4. We generalize the expression replacing the vector 3 2 4 with ω:

```
CM:⍉(⌽⍴POω)⍴¯1↓+\0,ω
```

```
        CM 3 2 4
0 0 0 0 0 0 0 0 0 0 0 0 0 0 0 0 0 0 0 0 0 0 0 0
3 3 3 3 3 3 3 3 3 3 3 3 3 3 3 3 3 3 3 3 3 3 3 3
5 5 5 5 5 5 5 5 5 5 5 5 5 5 5 5 5 5 5 5 5 5 5 5
```

The goal when writing algorithms is to have them depend upon only one or two array arguments. That way, you can use the result of one or more functions to produce new results which, in turn, can be used to produce additional results.

```
        'RPY∆□GBIW' POC 3 2 4
RRRRRRRRPPPPPPPPYYYYYYYY
∆∆∆∆□□□□∆∆∆∆□□□□∆∆∆∆□□□□
GBIWGBIWGBIWGBIWGBIWGBIW
```

When the defined function αPOCω executes, it invokes both CMω and POω:

```
POC:α[(POω)+CMω]
CM:⍉(⌽⍴POω)⍴¯1↓+\0,ω
PO:ωτιx/ω
```

We could have pursued a different strategy to create the same result that CMω generates. It serves to illustrate the versatility of APL and its "building-block" capabilities for expressing concepts. Let's examine the *product operator* represented by the . symbol. The product operator is dyadic and always takes a dyadic function, such as plus, as its right argument. When the left argument (typically another dyadic function) is omitted and represented instead by the null ∘ symbol, the resulting derived function ∘.g is called an *outer product*. The symbols ∘.+ indicate a *plus outer product*. The generalized expression α∘.+ω shows that this outer product applies to left and right array arguments.

44/ **Possible Outcomes as Characters**

For example, if α is the vector 1 3 and ω is the vector 0 4 8, the result is a matrix of sums:

```
      1 3∘.+0 4 8
1  5  9
3  7 11
```

The first row of the result is the sum of the first element in α and each element of ω. The second row is the sum of the second element in α and each element in ω, and so on.

The following sequence of APL expressions generates the same correction matrix we produced earlier. First, find the number of possible outcomes:

```
      ×/3 2 4
24
```

Then, find the values we need to add to the result of *PO* 3 2 4:

```
      ¯1↓+\0,3 2 4
0 3 5
```

Generate as many 0s as there are possible outcomes:

```
      (×/3 2 4)ρ0
0 0 0 0 0 0 0 0 0 0 0 0 0 0 0 0 0 0 0 0 0 0 0 0
```

Generate the correction matrix with the plus outer product:

```
      0 3 5∘.+(×/3 2 4)ρ0
0 0 0 0 0 0 0 0 0 0 0 0 0 0 0 0 0 0 0 0 0 0 0 0
3 3 3 3 3 3 3 3 3 3 3 3 3 3 3 3 3 3 3 3 3 3 3 3
5 5 5 5 5 5 5 5 5 5 5 5 5 5 5 5 5 5 5 5 5 5 5 5
```

We can now write:

```
      (¯1↓+\0,3 2 4)∘.+(×/3 2 4)ρ0
0 0 0 0 0 0 0 0 0 0 0 0 0 0 0 0 0 0 0 0 0 0 0 0
3 3 3 3 3 3 3 3 3 3 3 3 3 3 3 3 3 3 3 3 3 3 3 3
5 5 5 5 5 5 5 5 5 5 5 5 5 5 5 5 5 5 5 5 5 5 5 5
```

Thus, we could define an alternate correction function:

Possible Outcomes as Characters

```
CMALT:(¯1↓+\0,ω)∘.+(×/ω)⍴0

        CMALT 3 2 4
0 0 0 0 0 0 0 0 0 0 0 0 0 0 0 0 0 0 0 0 0 0 0 0
3 3 3 3 3 3 3 3 3 3 3 3 3 3 3 3 3 3 3 3 3 3 3 3
5 5 5 5 5 5 5 5 5 5 5 5 5 5 5 5 5 5 5 5 5 5 5 5
```

This chapter illustrates how to develop defined functions which allow us to take advantage of the power of the computer to do more and more of the work in solving a problem. These examples also demonstrate that great diversity in the approach to a problem is possible. In the preceding chapter, one of the exercises developed the numerical matrix of possible outcomes for dinners at a burger palace. We can now convert this numerical matrix to a character matrix using the function for changing possible outcomes to characters. We provide ten symbols and the vector of numbers of possible outcomes for each event:

```
     'FCRBPO=123'  POC 5 2 3
FFFFFFCCCCCCRRRRRRBBBBBBPPPPPP
OOO===OOO===OOO===OOO===OOO===
123123123123123123123123123123
```

The character vector has one symbol for fish, chicken, rib, burger, pizzaburger, potato salad, cole slaw, soda, tea and coffee respectively. The fifth column in the resulting matrix represents one possible dinner of fish, cole slaw and tea.

EXERCISES

6.1 Create a character matrix of all possible samples if one marble is drawn from a bowl containing an equal number of green and yellow marbles and then a second marble is drawn from a container of equal numbers of red, white and blue marbles.

6.2 Produce a character result of all possible samples similar to the one in 6.1, but draw first from the bowl of red, white and blue marbles and then from the bowl with green and yellow marbles.

6.3 Produce a character matrix of the possible outcomes when a coin and a die are tossed together.

6.4 Suppose a traffic light is either red or green, and a driver can go left, straight or right. Assume that another light is either red or green, and the driver once again can go left, straight or right. Produce a character matrix of all 36 different sequences of events.

6.5 Create a 2-row, 52-column character array representing a conventional bridge deck of cards.

7/ ALL POSSIBLE COMBINATIONS

$$AC:(((2\times\rho\omega)\rho0\ 1)\backslash\omega)POC(\rho\omega)\rho2$$
$$\alpha\backslash\omega$$
$$\alpha\star\omega$$

As you walk into the supermarket, you find volunteers taking up a collection for the local rescue squad. In your pocket you have a penny, a nickel, a dime and a quarter. What are all the different possible contributions you might make?

Notice that there is a subtle change in the key issue here. We are not interested in *which* of the four coins to donate. Instead, we must decide whether or not to donate each of the four coins. That is, there are two possibilities for each event: "yes" or "no;" "success" or "failure;" 1 or 0. When there are only two possibilities for an event, it is termed *binomial*. Let P stand for donating the penny and $\underline{P}$ represent the alternative choice—not giving the penny. We can designate the choices for the nickel, dime and quarter in a similar fashion:

```
     'PPNNDDQQ' POC 2 2 2 2
PPPPPPPPPPPPPPPP
NNNNNNNNNNNNNNNN
DDDDDDDDDDDDDDDD
QQQQQQQQQQQQQQQQ
```

If we choose a blank to indicate that a coin is not donated, we have a more readable result:

```
     ' P N D Q' POC 2 2 2 2
        PPPPPPPP
    NNNN    NNNN
  DD  DD  DD  DD
 Q Q Q Q Q Q Q Q
```

The displayed characters indicate all possible donations. Putting it another way, the matrix shows all of the possible combinations of 0, 1, 2, 3 and 4 coins in its 16 columns. The concept of all possible combinations is fundamental to an understanding of probability theory.

We can define an *all combinations* function $AC\omega$ to produce a character array illustrating all combinations of a series of choices. One of the best ways to develop skill in the use of any language is to "read" successively complex expressions. As we define and analyze $AC\omega$ and other functions, we will learn how to take advantage of the APL computer's capability:

      $AC:(((2\times\rho\omega)\rho0\ 1)\backslash\omega)POC(\rho\omega)\rho2$

```
          AC 'PNDQ'
             PPPPPPPP
         NNNN    NNNN
       DD  DD  DD  DD
      Q Q Q Q Q Q Q Q
```

$AC\omega$ generates the right and left arguments of $\alpha POC\omega$, evaluates that expression and displays its result. That is, by defining $AC\omega$, we are exploiting the power of the APL computer to derive the appropriate arguments to POC for us. Defining functions in this fashion aids in problem solving *and* it saves us time by freeing us from typing in long expressions each time we need to perform a particular calculation.

Reading the definition of $AC\omega$ from the right, the expression $(\rho\omega)\rho2$ creates the right argument for POC. The expression in parentheses $(\rho\omega)$ is evaluated as a unit:

```
          ρ'PNDQ'
      4
```

When we find all possible combinations, each event must have only two alternatives. We use the reshape function to generate a vector of one 2 for each event:

```
          4ρ2
      2 2 2 2
```

This is the desired right argument for POC.

The entire expression $(((2\times\rho\omega)\rho0\ 1)\backslash\omega)$ generates the left argument of POC. In the case of our example, we need the 8-element character vector ' P N D Q '. The expression $2\times\rho\omega$ in the innermost set of parentheses is evaluated first. It computes twice the number of elements in ω:

```
          2×ρ'PNDQ'
      8
```

Working outward, the entire expression in the next set of parentheses is evaluated. $(2\times\rho\omega)\rho0\ 1$ generates a vector of alternating 0s and 1s:

```
      8ρ0 1
0 1 0 1 0 1 0 1
```

Any array containing only 0s and 1s is called a Boolean array. The Boolean vector in the preceding example is an appropriate left argument to use with the *expand* function α\ω. There must be as many 1s in the vector in α as there are elements in the vector in ω:

```
      0 1 0 1 0 1 0 1\'PNDQ'
 P N D Q
```

The result of expanding a character vector contains a blank for each position in which a 0 appears in the Boolean vector. The result in our example begins with a blank and has eight elements. The shape of a character array is sometimes not apparent because of the blanks which it contains. The shape function allows us to check that the created array is the one we intended:

```
      ρ((2×ρ'PNDQ')ρ0 1)\'PNDQ'
8
```

This last step is not part of the definition of *AC*, but we can use it to reassure ourselves that we're proceeding properly while we're developing the definition.

Continuing with the example, the entire expression

```
((2×ρω)ρ0 1)\ω
```

in the outermost set of parentheses generates the left argument for *POC*:

```
      ((2×ρ'PNDQ')ρ0 1)\'PNDQ'
 P N D Q
```

Now, let's execute the entire expression:

```
      (((2×ρ'PNDQ')ρ0 1)\'PNDQ')POC(ρ'PNDQ')ρ2
         PPPPPPPP
      NNNN    NNNN
   DD  DD  DD  DD
  Q Q Q Q Q Q Q Q
```

We have produced a result from a single argument. That is, by supplying just `'PNDQ'`, we can find all the donations possible from our pocket change of a penny, nickel, dime and quarter. We can generalize this expression and define an *all combinations* function to produce all of the possible combinations of the elements in ω:

$AC:(((2\times\rho\omega)\rho 0\ 1)\backslash\omega)POC(\rho\omega)\rho 2$

Now, we need only enter AC `'PNDQ'` to get our result:

```
    AC 'PNDQ'
       PPPPPPPP
   NNNN    NNNN
 DD  DD  DD  DD
Q Q Q Q Q Q Q Q
```

This algorithm is designed to take a vector of character elements and generate a matrix showing all of the possible combinations of the elements.

Defined functions are used to solve similar problems with different sets of data. Suppose you go to get a pizza and find that sausage, mushrooms, anchovies, peppers and onions are all optional items. What are all the possible combinations and how many choices are there?

```
      AC 'SMAPO'
            SSSSSSSSSSSSSSSS
       MMMMMMMM        MMMMMMMM
    AAAA    AAAA    AAAA    AAAA
  PP  PP  PP  PP  PP  PP  PP  PP
 O O O O O O O O O O O O O O O O
```

Notice that the first column indicates the rejection of all five optional items because it contains only blanks. The second column represents choosing only onions. The shape of the result is 5 rows and 32 columns:

```
     ρAC 'SMAPO'
5 32
```

Thus, 32 different pizzas can be made from the 5 different optional items. We could get this number another way:

```
     ρ'SMAPO'
5
```

52/ All Possible Combinations

```
      5⍴2
2 2 2 2 2

      ×/2 2 2 2 2
32
```

We could also use the *power* function α⋆ω to calculate the number of possible combinations. The base is α and the exponent is ω:

```
      2⋆5
32
```

Do you remember the random walker who walked ten city blocks? The number of different possible walks is:

```
      4⋆10
1048576
```

EXERCISES

7.1 Alex, Ben, Cara, Dan and Eloise are all invited to a class party. They each decide independently if they will attend. Produce a character matrix that shows all of the combinations of who might attend.

7.2 How many different combinations are possible in **7.1**?

7.3 Suppose Fred also gets an invitation to the party. Create a new character matrix to include him.

7.4 How many different combinations are possible in **7.3**?

7.5 George has homework in English, social studies, art and math. Whether or not he does the assignment for each course is up to George. Create a character matrix to show all of the possible combinations of courses in which he might be prepared.

7.6 How many different combinations are possible in **7.5**?

8/ PROBABILITY

$$f \neq \omega$$
$$+ \neq \omega$$
$$\alpha = \omega$$
$$\alpha \neq \omega$$
$$, \omega$$
$$\alpha / \omega$$
$$\alpha \iota \omega$$
$$\blacktriangle \omega$$
$$NUB:NP,\omega$$
$$NP:((\omega\iota\omega)=\iota\rho\omega)/\omega$$
$$ONUB:(NUB\omega)[\blacktriangle NUB\omega]$$
$$FT:FP,\omega$$
$$FP:(ONUB\omega),[.5]+/(ONUB\omega)\circ .=\omega$$
$$\alpha\circ .=\omega$$
$$+/\omega$$
$$f[\omega]$$
$$\alpha,[X]\omega$$

What is the probability of obtaining 0, 1, 2, 3 or 4 heads when 4 coins are tossed? First, we need a definition of *probability*. It is actually a very simple notion: a probability is the number of "successful" outcomes divided by the number of possible outcomes. Earlier, when we developed tree diagrams and the matrix of possible outcomes for tossing four coins, we were interested in all possibilities. Now, we are focusing our attention on only the "successes." We can think of a success as a head, an H or a 1:

```
    ' H'[PO 2 2 2 2]
       HHHHHHHH
    HHHH    HHHH
   HH HH  HH HH
   H H H H H H H H
```

There are 16 possible outcomes since each coin has 2 options:

```
       2*4
16
```

When we examine the columns of the possible outcomes matrix for the number of times 3 heads occur, we find that 4 of the 16 meet the

Probability /55

condition. The probability of obtaining three heads is this ratio:

```
        4÷16
0.25
```

Similarly, only one column shows four *H*s, so the probability of tossing four heads is:

```
        1÷16
0.0625
```

Note that the probability of tossing no heads is also:

```
        1÷16
0.0625
```

Since for any given toss of 4 coins there are always 16 possible outcomes, 16 is the divisor for calculating the probability of tossing a given number of heads.

By definition, a probability is always a number from 0 to 1. A probability of 0 describes a completely improbable outcome such as getting five heads when four coins are tossed. A probability of 1 describes the probability of complete success, such as drawing a penny from a pocket that has nothing but pennies in it. In the first case, success is never expected and in the latter, success is always expected.

The numerical matrix of indexes generated by $PO\omega$ contains a 1 for each head:

```
      PO 2 2 2 2
0 0 0 0 0 0 0 0 1 1 1 1 1 1 1 1
0 0 0 0 1 1 1 1 0 0 0 0 1 1 1 1
0 0 1 1 0 0 1 1 0 0 1 1 0 0 1 1
0 1 0 1 0 1 0 1 0 1 0 1 0 1 0 1
```

A *reduction along the first axis* is represented by the notation $f\!\not/\omega$. When f is the plus function, $+\!\not/\omega$ gives the sum of each column of a matrix in ω. In our example, the sums represent the number of possible heads in any given toss:

```
     +/PO 2 2 2 2
0 1 1 2 1 2 2 3 1 2 2 3 2 3 3 4
```

To have the computer find out how many times three heads occur, we

56/ Probability

can use the *equals* function α=ω to produce a Boolean vector:

```
      3=0 1 1 2 1 2 2 3 1 2 2 3 2 3 3 4
0 0 0 0 0 0 0 1 0 0 0 1 0 1 1 0
```

A 1 in the result means that there is a 3 in the corresponding position in the right argument:

```
      3=+/PO 2 2 2 2
0 0 0 0 0 0 0 1 0 0 0 1 0 1 1 0
```

The sum of the Boolean result of 3=+/PO 2 2 2 2 gives the number of times three heads occur in the possible outcomes matrix:

```
      +/3=+/PO 2 2 2 2
4
```

Another aspect of a probability is that once the definition of a success is formulated, all other possible outcomes for the trial are *not* successes:

```
      +/3≠+/PO 2 2 2 2
12

      +/PO 2 2 2 2
0 1 1 2 1 2 2 3 1 2 2 3 2 3 3 4

      3≠0 1 1 2 1 2 2 3 1 2 2 3 2 3 3 4
1 1 1 1 1 1 1 0 1 1 1 0 1 0 0 1

      +/3≠0 1 1 2 1 2 2 3 1 2 2 3 2 3 3 4
12

      12÷16
0.75
```

The sum of the successful and unsuccessful outcomes in a given situation is always 1.

```
      +/4 12÷16
1
```

We can find the number of times a series of successes can occur for a particular event by using a more general approach. That is, we can

Probability /57

define a function which directly creates a frequency table showing the number of times the elements of any array occur in the array. For example,

$$+\neq PO\ 2\ 2\ 2\ 2$$
0 1 1 2 1 2 2 3 1 2 2 3 2 3 3 4

$$FT \leftarrow \neq PO\ 2\ 2\ 2\ 2$$
0 1
1 4
2 6
3 4
4 1

The first column in the preceding result shows the number of successes or heads—0, 1, 2, 3 or 4. The second column contains the number of respective times we can expect that number of heads when tossing four coins. By inspecting the result of ' H ' $[PO\ 2\ 2\ 2\ 2]$ we could determine that no heads are possible once, one head is possible four times, two heads are possible six times, three are possible four times and four heads are possible once. Defining $FT\omega$ is obviously an easier way of achieving our goal. Let's examine the development of a function which lets the computer do most of the drudgery.

First, let's define a pair of functions NUB and NP to sort through the elements of any array and produce a list of the unique elements contained in the array:

$NUB:NP,\omega$
$NP:((\omega\iota\omega)=\iota\rho\omega)/\omega$

$\qquad NUB\ 2\ 4\rho 6\ 3\ 8\ 8\ 5$
6 3 8 5

The *ravel* function $,\omega$ in NUB changes any array in ω into a vector:

$\qquad 2\ 4\rho 6\ 3\ 8\ 8\ 5$
6 3 8 8
5 6 3 8

$\qquad ,2\ 4\rho 6\ 3\ 8\ 8\ 5$
6 3 8 8 5 6 3 8

The vector result 6 3 8 8 5 6 3 8 is the argument of the *nub proce-*

58/ **Probability**

dure function *NP*. We insure, by defining two functions, that the argument of *NP* is always a vector when it is used in *NUB*.

The defined function *NP* makes use of the *replicate* function α/ω which is very similar to expand. In order to understand how replicate works, we need to examine its arguments. In our example, the right argument of replicate is the vector 6 3 8 8 5 6 3 8. Its left argument (ω⍳ω)=⍳ρω requires several steps to generate:

```
        ρ6 3 8 8 5 6 3 8
8

        ⍳8
0 1 2 3 4 5 6 7
```

The expression ⍳ρω produces a vector which can be viewed as numbering the elements of the array in ω. The first element in the result is ⎕IO.

Next, *NP* uses the equals function to compare the result of ω⍳ω with the numbering vector 0 1 2 3 4 5 6 7. The *index of* function α⍳ω gives the least index in α of each element in ω. That is, the result of the index of function shows the *first* position in α at which the corresponding element of ω is found:

```
        6 3 8 8 5 6 3 8⍳6 3 8 8 5 6 3 8
0 1 2 2 4 0 1 2
```

If the first position of an element in the result of the index of function is equal to its corresponding element in the numbering vector, that element is a first occurrence. Finding the first occurrence of any element in an array is a way of identifying each unique element in it:

```
        0 1 2 2 4 0 1 2=0 1 2 3 4 5 6 7
1 1 1 0 1 0 0 0
```

Now that we've generated the arguments to replicate, let's examine how this function works. As mentioned earlier, replicate is similar to expand. The value in each position of the axis of ω along which replicate is applied is reproduced as many times as indicated by the value of the corresponding element in α. When α is Boolean, replicate selects the elements in ω where there is a 1 in α. It deletes elements, rather than expands, whenever there is a 0 in α:

Probability /59

```
      1 1 1 0 1 0 0 0/6 3 8 8 5 6 3 8
6 3 8 5
```

Each element in this vector result is unique.

```
      NUB 2 4ρ6 3 8 8 5
6 3 8 5
```

Arranging the elements in the result of *NUB*ω in order—from smallest to largest—will make the analysis of this data simpler. The defined function *ordered nub ONUB*ω does this:

ONUB:(NUBω)[⍋NUBω]

The *upgrade* function ⍋ω returns a vector of the indexes of the smallest to the largest values in ω:

```
      ⍋6 3 8 5
1 3 0 2
```

As we have seen, the index function α[ω] selects the elements in α according to the index values in ω. The array in α can be either character or numeric:

```
      NUB 2 4ρ6 3 8 8 5
6 3 8 5

      6 3 8 5[1 3 0 2]
3 5 6 8
```

The ordered nub gives a vector of the unique values in ω arranged in ascending order:

```
      ONUB 2 4ρ6 3 8 8 5
3 5 6 8
```

When the array in ω contains hundreds or thousands of elements, the utility of *ONUB*ω is quite obvious.

Now we can define a pair of functions to generate a *frequency table*:

FT:FP,ω
FP:(ONUBω),[.5]+/(ONUBω)∘.=ω

As before, one of the functions (FT) simply insures that the argument of the other (FP) is a vector.

In the *frequency procedure* function FP, the *equals outer product* denoted by α∘.=ω compares the result of $ONUB$ω with the raveled array in ω:

```
      3 5 6 8∘.=6 3 8 8 5 6 3 8
0 1 0 0 0 0 1 0
0 0 0 0 1 0 0 0
1 0 0 0 0 1 0 0
0 0 1 1 0 0 0 1
```

This is much like the process of "tallying." For example, examine the third row of the result. A 1 appears wherever the 6 in α is equal to the corresponding value in ω. That is, the third element in α is compared to each element in α. A 1 in the third row means that the equals relationship is true for it and the corresponding element in ω. A 0 means the relationship is false.

A *plus reduction along the last axis* +/ω gives the sum of the elements in each row of the matrix created by the equals outer product:

```
      +/3 5 6 8∘.=6 3 8 8 5 6 3 8
2 1 2 3
```

The resulting vector represents the relative frequency with which the elements in 3 5 6 8 appear in 6 3 8 8 5 6 3 8. For example, 8 appears three times.

The notation $f[\omega]$ denotes the *axis operator*, and it explicitly indicates the axis along which the function in f is to be applied. The axis operator is subject to the value of □IO. That is, in origin 0, a 3-axis array has axis numbers of 0, 1 and 2.

The expression α,[X]ω indicates that the arrays α and ω are to be joined along the axis specified by X. When the value of X is an integer indicating one of the existing axes of α or ω, α,[X]ω is the notation for the catenate function, one with which you're already familiar. When X is *not* an integer, the axis operator indicates the *new* axis along which the two arrays will be joined and α,[X]ω represents the *laminate* function. In the definition of FP, the arrays to be joined are both vectors—1-axis arrays. Putting .5 (or any decimal between 0 and 1 when □IO is 0) makes the new axis the second axis and the result of the lamination is a 4-by-2 matrix. (If X were a decimal between ¯1 and 0, the new axis would have been the first axis, and the result would have been a 2-by-4 matrix.)

Laminating along the second axis creates a very readable result. The ordered nub 3 5 6 8 in α is the first column of the matrix and the vector of the relative frequencies 2 1 2 3 in ω is the second column:

```
        FT 2 4ρ6 3 8 8 5
3 2
5 1
6 2
8 3
```

With these five algorithms, we can generate a frequency table of the number of times 0, 1, 2, 3 and 4 heads occur in the matrix of possible outcomes for tossing 4 coins:

```
FT:FP,ω
FP:(ONUBω),[.5]+/(ONUBω)∘.=ω
ONUB:(NUBω)[⍋NUBω]
NUB:NP,ω
NP:((ωιω)=ιρω)/ω

    +/PO 2 2 2 2
0 1 1 2 1 2 2 3 1 2 2 3 2 3 3 4

    FT+/PO 2 2 2 2
0 1
1 4
2 6
3 4
4 1
```

The expression +/(ONUBω)∘.=ω produces the frequencies 1 4 6 4 1 from the vector argument +/PO2 2 2 2. The sum of the frequencies is also the total number of possible outcomes:

```
    +/1 4 6 4 1
16
```

The respective probabilities of obtaining 0, 1, 2, 3 and 4 heads are found by division:

```
       1 4 6 4 1÷16
0.0625 0.25 0.375 0.25 0.0625
```

The sum of these probabilities—which encompass all the possible outcomes—is 1:

```
      +/1 4 6 4 1÷16
1
```

We can simulate 160 tosses of 4 pennies for comparison, and then use $FT\omega$ to get a frequency table of the result:

```
      FT+/?4 160ρ2
0  6
1 46
2 64
3 38
4  6
```

Tossing 4 coins 160 times is an example of a binomial experiment. The basis for binomial experiments is the Bernoulli trial, in which an outcome of a trial is either a success or a failure. The chance of success is the same for every trial. The experiment consists of a fixed number of repeated trials—in this case 4—which are independent events.

Based on the table of relative frequencies, for 160 tosses we expect 10 times as many elements in each category or:

```
         10×1 4 6 4 1
10 40 60 40 10
```

An analysis of how close a simulation is to the theoretical expectation leads to a study of statistics.

EXERCISES

8.1 Find the probabilities for obtaining 0, 1, 2 and 3 heads when 3 pennies are tossed.

8.2 Simulate 800 tosses of 3 pennies and use the defined functions necessary to obtain a frequency table of your result.

8.3 What are the expected number of times you'd anticipate getting 0, 1, 2 and 3 heads when 3 pennies are tossed 800 times?

8.4 Find the probabilities for obtaining 0, 1, 2, 3, 4 and 5 heads when 5 pennies are tossed.

8.5 Simulate 3200 tosses of 5 pennies and generate a frequency table of your result.

8.6 What are the expected number of times you'd anticipate getting 0, 1, 2, 3, 4 and 5 heads when 5 pennies are tossed 3200 times?

8.7 Define another alternative for the correction matrix functions *CM* and *CMALT* developed in the "Possible Outcomes as Characters" chapter. Use the replicate function and call your new function *CMALT*2.

9/ BINOMIAL DISTRIBUTIONS

$$\alpha!\omega$$
$$\alpha\circ.!\omega$$
$$PT:(\iota 1+\omega)\circ.!\iota 1+\omega$$
$$\alpha\lceil\omega$$
$$\lceil/\omega$$
$$\alpha\circ.>\omega$$
$$HIST:' \quad \square'[((\iota 1+\omega)!\omega)\circ.>\iota\lceil/(\iota 1+\omega)!\omega]$$
$$FTN:FTNP,\omega$$
$$FTNP:(NUB\omega),[.5]+/(NUB\omega)\circ.=\omega$$
$$BP:((\iota 1+\alpha)!\alpha)\times(\omega*\iota 1+\alpha)\times(1-\omega)*\phi\iota 1+\alpha$$

What is the distribution of the probabilities of the combinations of 0, 1, 2, 3, 4 or 5 heads when 5 coins are tossed?

```
        ' H'[PO 2 2 2 2 2]
                HHHHHHHHHHHHHHHH
         HHHHHHHH        HHHHHHHH
    HHHH       HHHH    HHHH      HHHH
  HH  HH  HH  HH   HH  HH   HH  HH
 H H H H H H H H H H H H H H H H
```

A frequency table of the sums of the columns of $PO\omega$ is:

```
        FT+/PO 2 2 2 2 2
0   1
1   5
2  10
3  10
4   5
5   1
```

This is a theoretical distribution of the frequency of occurrence of the number of heads shown in the columns of the possible outcomes matrix.

Binomial Distributions

The third row, 2 10, indicates that there are 10 possible outcomes for 2 heads when 5 coins are tossed. The *binomial coefficients* function α!ω gives the same result when α is 2 and ω is 5:

```
      2!5
10
```

The binomial coefficients function provides the number of distinct ways α items can be selected from a set of ω items.

We can find the entire frequency distribution by entering all the possible numbers of heads in α:

```
      0 1 2  3  4 5!5
1 5 10 10 5 1
```

The familiar Pascal triangle can be obtained with α∘.!ω, the *binomial combinations outer product*:

```
          (ι6)∘.!ι6
1 1 1 1 1  1
0 1 2 3 4  5
0 0 1 3 6 10
0 0 0 1 4 10
0 0 0 0 1  5
0 0 0 0 0  1
```

We can generalize this expression and define a function to produce binomial coefficients in a Pascal triangle:

PT:(ι1+ω)∘.!ι1+ω

```
         PT 5
1 1 1 1 1  1
0 1 2 3 4  5
0 0 1 3 6 10
0 0 0 1 4 10
0 0 0 0 1  5
0 0 0 0 0  1
```

We add 1 to the arguments of the binomial combinations outer product because we're working in origin 0.

The sum of each column in the result of PTω is:

```
      +/PT 5
1 2 4 8 16 32
```

The final element in the result represents the total number of possible outcomes for tossing 5 coins. That is, there are 32 possible combinations.

We can compute probability distributions from frequency distributions by making a matrix of appropriate divisors for the values in each column of the result of $PT\omega$:

```
      6 6ρ+/PT 5
1 2 4 8 16 32
1 2 4 8 16 32
1 2 4 8 16 32
1 2 4 8 16 32
1 2 4 8 16 32
1 2 4 8 16 32
```

Before dividing, we can change the value of the system variable *print precision* $\Box PP$ to 4:

```
      □PP←4
```

The APL system displays decimal numbers with 4 significant digits when $\Box PP$ is set to 4:

```
  (PT 5)÷6 6ρ+/PT 5
1      0.5      0.25     0.125    0.0625   0.03125
0      0.5      0.5      0.375    0.25     0.1563
0      0        0.25     0.375    0.375    0.3125
0      0        0        0.125    0.25     0.3125
0      0        0        0        0.0625   0.1563
0      0        0        0        0        0.03125
```

The final column is the distribution of probabilities for obtaining 0, 1, 2, 3, 4 or 5 heads when 5 coins are tossed. The fifth column contains the same probability distribution for tossing four coins that we obtained in the preceding chapter.

We can "look" at the frequency of throwing 0, 1, 2, 3, 4 or 5 heads when tossing 5 coins by creating a histogram. This is a useful exercise because it is sometimes easier to understand the meaning of numbers graphically.

```
      HIST 5
□
□□□□
□□□□□□□□□□
□□□□□□□□□□
□□□□□
□
```

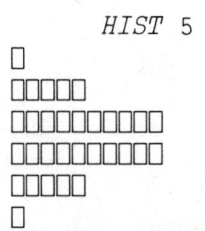

First, we need the frequency distribution:

```
      (ι1+5)!5
1 5 10 10 5 1
```

By inspection, we can see that 10 is the largest number in the vector result. The APL *maximum* function $\alpha \lceil \omega$ gives the larger element of α or ω:

```
      5⌈1
5
```

The *maximum reduction* $\lceil /\omega$ instructs the APL computer to find the largest element in the distribution frequency vector for us:

```
      ⌈/1 5 10 10 5 1
10
```

The reduction operator applies the maximum function between each element of the argument. Thus, in the example, 5⌈1 is evaluated first. The *result* of that expression, 5, is then compared with the next element in ω, the 10. The result of 10⌈5 is compared with the following element, and so on.

The maximum reduction of the frequency distribution vector gives us the number of positions we need to represent the largest frequency. That is, our bar graph should show 10 □s for throwing 2 and 3 heads respectively. We can also create the index values for each of those positions by using the index generator:

```
      ι⌈/(ι1+5)!5
0 1 2 3 4 5 6 7 8 9
```

Now we can gain some insight into the relationship of the values in the frequency distribution vector to the values of the indices necessary to represent its largest element graphically. We can also use the

68/ Binomial Distributions

power of APL to operate on arrays of data:

```
        1 5 10 10 5 1∘.>0 1 2 3 4 5 6 7 8 9
1 0 0 0 0 0 0 0 0 0
1 1 1 1 1 0 0 0 0 0
1 1 1 1 1 1 1 1 1 1
1 1 1 1 1 1 1 1 1 1
1 1 1 1 1 0 0 0 0 0
1 0 0 0 0 0 0 0 0 0
```

The *greater than outer product* denoted by α∘.>ω creates a Boolean matrix which bears a striking similarity to the histogram we seek to generate. Each element in α is compared with every element in ω. A 1 appears in the result whenever the element in α is greater than the element in ω. Note that the length of α and ω do not have to match. In the example, the length of α determines the number of rows in the result. Finally, we can use the Boolean matrix to index the 2-element character vector ' □' to create the histogram:

```
        ' □'[((ι1+5)!5)∘.>ι⌈/(ι1+5)!5]
□
□□□□□
□□□□□□□□□□
□□□□□□□□□□
□□□□□
□
```

HIST:' □'[((ι1+ω)!ω)∘.>ι⌈/(ι1+ω)!ω]

The frequency distribution of possible outcomes for tossing 7 coins is represented by the following histogram. Each □ stands for one of the 2*7 or 128 total possible outcomes of 0, 1, 2, 3, 4, 5, 6 or 7 heads:

```
        HIST 7
□
□□□□□□□
□□□□□□□□□□□□□□□□□□□□□
□□□□□□□□□□□□□□□□□□□□□□□□□□□□□□□□□□□
□□□□□□□□□□□□□□□□□□□□□□□□□□□□□□□□□□□
□□□□□□□□□□□□□□□□□□□□□
□□□□□□□
□
```

Binomial Distributions /69

Let's find all the combinations of the letters in the word *DANCER*:

```
AC 'DANCER'
                                DDDDDDDDDDDDDDDDDDDDDDDDDDDDDDDD
            AAAAAAAAAAAAAAAA                AAAAAAAAAAAAAAAA
    NNNNNNNN         NNNNNNNN        NNNNNNNN        NNNNNNNN
  CCCC    CCCC    CCCC    CCCC    CCCC    CCCC    CCCC    CCCC
 EE  EE  EE  EE  EE  EE  EE  EE  EE  EE  EE  EE  EE  EE  EE  EE
R R R R R R R R R R R R R R R R R R R R R R R R R R R R R R R R
```

Since the possible outcomes matrix generated by *POω* lies at the heart of the function *ACω*, the frequency distribution of 0, 1, 2, 3, 4, 5 or 6 letters is 1, 6, 15, 20, 15, 6 or 1—the binomial coefficients:

```
      (ι7)!6
1 6 15 20 15 6 1
```

If we assume that selecting or not selecting a letter is equally likely, we obtain the binomial probability distribution:

```
        (ι7),[.5]((ι7)!6)÷2*6
0        0.01563
1        0.09375
2        0.2344
3        0.3125
4        0.2344
5        0.09375
6        0.01563
```

It is worth noting that a particular combination of letters from the word *DANCER* is not considered to be different from another combination containing the same letters. That is, from the standpoint of combinations, *ACE* does not differ from *EAC*. The order of the elements is not important. The study of the number of ways the different letters can be arranged is included in a later chapter about permutations.

If pies for a picnic basket are chosen by selecting or not selecting—with equal likelihood—from peach, apple, banana cream, strawberry, cherry, raspberry and blueberry, what is the probability of selecting exactly three of the seven different pies?

```
       3!7
35
```

Thirty-five different combinations of three different pies can be chosen.

70/ **Binomial Distributions**

The number of all possible combinations is:

      2*7
128

Thus, the probability is:

      (3!7)÷2*7
0.2734

This result indicates that approximately 27 percent of the time, the picnic basket is likely to contain exactly 3 different pies.

Suppose we are interested in the probability distribution for a basketball player obtaining 0, 1, 2 or 3 successes in 3 free throws, when the probability of success on any one free throw is .25. Here we have a binomial situation where the probability of failure or success is not the same. We begin as before with a matrix of possible outcomes:

      PO 2 2 2
0 0 0 0 1 1 1 1
0 0 1 1 0 0 1 1
0 1 0 1 0 1 0 1

Next we use these values to index the numeric vector .75 .25 representing the probability of failure and the probability of success. The array entered in α of α[ω] can be either character or numeric:

      .75 .25[PO 2 2 2]
0.75 0.75 0.75 0.75 0.25 0.25 0.25 0.25
0.75 0.75 0.25 0.25 0.75 0.75 0.25 0.25
0.75 0.25 0.75 0.25 0.75 0.25 0.75 0.25

Next we find the product of the elements in each column:

      ×/.75 .25[PO 2 2 2]
0.4219 0.1406 0.1406 0.04688 0.1406 0.04688
      0.04688 0.01563

The second column in the matrix created by .75 .25[PO 2 2 2] represents a miss, a miss and a basket. The third column shows a miss, a basket and a miss. The product of the values in each column is the same; that is, 0.1406. We can define a pair of functions to generate a *frequency table not ordered* of these probabilities:

```
      FTN:FTNP,ω
      FTNP:(NUBω),[.5]+/(NUBω)∘.=ω

            FTN x≠.75 .25[PO 2 2 2]
      0.4219    1
      0.1406    3
      0.04688   3
      0.01563   1
```

The product of the rows of this matrix, the probability of each possible outcome times the frequency of its occurrence, gives the binomial probability distribution:

```
            x/FTN x≠.75 .25[PO 2 2 2]
      0.4219 0.4219 0.1406 0.01563
```

A more traditional way of obtaining the *binomial probabilities* is to generalize the expression with α representing the number of trials and ω the probability of success for each:

```
      BP:((ι1+α)!α)×(ω*ι1+α)×(1-ω)*⌽ι1+α
```

This represents the product of the following three vectors when α is 3 and ω is .25:

```
            .75*⌽ι4
      0.4219 0.5625 0.75 1

            .25*ι4
      1 0.25 0.0625 0.01563

            (ι4)!3
      1 3 3 1

            3 BP .25
      0.4219 0.4219 0.1406 0.01563
```

The third element represents the three probabilities which contain one miss and two baskets. We can use the laminate function to present this information in a table.

	(ιρ3 BP .25),[.5]3 BP .25
0	0.4219
1	0.4219
2	0.1406
3	0.01563

We interpret the third row to mean that a basketball player with a probability of success of .25 on an individual throw is likely to get two successes in three throws 14.06 percent of the time.

EXERCISES

9.1 What is the probability that two coins will be donated to the rescue squad from the loose change of a penny, nickel, dime and quarter? Assume that each coin is equally likely to be given or not given.

9.2 What is the probability that exactly three options will be selected when sausage, mushrooms, anchovies, peppers and onions are available to be added to a regular pizza? Assume that each choice is equally likely to be selected or not selected.

9.3 Find the matrix of all the combinations of letters in the word *SHARED*.

9.4 Find the probability distribution for 9.3.

9.5 Produce a histogram for 9.3.

9.6 What is the probability that a combination of letters from the word *SHARED* has four letters?

9.7 It is known that 60 percent of mice inoculated with a serum are protected from a certain disease. Create the binomial distribution table of the theoretical probabilities for the numbers of disease-free mice in samples of 5 mice.

9.8 Simulate 100 samples of the 5 mice described in the preceding example and produce a frequency table of the sampling distribution.

10/ COMBINATIONS

$$CR: \emptyset(\alpha=+\neq PO\omega\rho 2)/PO\omega\rho 2$$
$$COMB: \emptyset((\alpha!\omega),\alpha)\rho(,\alpha CR\omega)/,(\rho\alpha CR\omega)\rho\iota\omega$$

Suppose an Airedale, beagle, corgi and dachshund all live on the same street. Their four owners are as likely as not to walk them before breakfast. Our next goal is to produce a matrix showing all of the combinations for exactly two of the four dogs being walked before breakfast on a given morning. From the preceding chapter, we know that the number of combinations is:

```
       2!4
6
```

And, we can get all combinations:

```
     AC 'ABCD'
       AAAAAAAA
    BBBB    BBBB
 CC  CC  CC  CC
D D D D D D D D
```

We are interested in selecting only those columns which represent two dogs being taken for a walk. The defined function *combinations as rows* produces the needed subset of the possible outcomes matrix and provides the combinations as rows of a matrix:

$CR: \emptyset(\alpha=+\neq PO\omega\rho 2)/PO\omega\rho 2$

Using 4 as ω creates the vector 2 2 2 2. The possible outcomes matrix is generated next:

```
    PO 2 2 2 2
0 0 0 0 0 0 0 0 1 1 1 1 1 1 1 1
0 0 0 0 1 1 1 1 0 0 0 0 1 1 1 1
0 0 1 1 0 0 1 1 0 0 1 1 0 0 1 1
0 1 0 1 0 1 0 1 0 1 0 1 0 1 0 1
```

A Boolean selection vector is needed again:

```
        2=+/PO 4ρ2
0 0 0 1 0 1 1 0 0 1 1 0 1 0 0 0
```

The replicate function applied along the last axis of the result of
PO 4ρ2 deletes the columns whose sums are *not* 2:

```
       (2=+/PO 4ρ2)/PO 4ρ2
0 0 0 1 1 1
0 1 1 0 0 1
1 0 1 0 1 0
1 1 0 1 0 0
```

Finally, the transpose function interchanges the rows and columns:

```
       ⌽(2=+/PO 4ρ2)/PO 4ρ2
0 0 1 1
0 1 0 1
0 1 1 0
1 0 0 1
1 0 1 0
1 1 0 0
```

As was the case earlier when we developed a strategy to use the possible outcomes matrix to index a character vector argument, the result of α*CR*ω needs to have the 1s in the second column replaced by 2s, and the 1s in the third and fourth columns replaced by 3s and 4s respectively. The defined function α*COMB*ω generates the index values necessary to produce all of the *combinations* as a character matrix:

COMB:⌽((α!ω),α)ρ(,α*CR*ω)/,(ρα*CR*ω)ρɩω

The shape of the matrix produced by α*CR*ω establishes the shape of a matrix of index values:

```
     2 CR 4
0 0 1 1
0 1 0 1
0 1 1 0
1 0 0 1
1 0 1 0
1 1 0 0
```

```
      ⍴2 CR 4
6 4
```

```
      6 4⍴⍳4
0 1 2 3
0 1 2 3
0 1 2 3
0 1 2 3
0 1 2 3
0 1 2 3
```

The two matrices are each raveled and the replicate function deletes the unnecessary elements:

```
      ,(⍴2 CR 4)⍴⍳4
0 1 2 3 0 1 2 3 0 1 2 3 0 1 2 3 0 1 2 3 0 1 2 3
```

```
      ,2 CR 4
0 0 1 1 0 1 0 1 0 1 1 0 1 0 0 1 1 0 1 0 1 1 0 0
```

```
      (,2 CR 4)/,(⍴2 CR 4)⍴⍳4
2 3 1 3 1 2 0 3 0 2 0 1
```

These elements are formed into a matrix of six rows and two columns. The number of rows, 6, represents the number of combinations of chosen items and the number of columns, 2, represents the number of items chosen:

```
      ((2!4),2)⍴(,2 CR 4)/,(⍴2 CR 4)⍴⍳4
2 3
1 3
1 2
0 3
0 2
0 1
```

We use transpose here simply to allow us to continue to think of events as columns in a matrix:

```
      ⌽((2!4),2)⍴(,2 CR 4)/,(⍴2 CR 4)⍴⍳4
2 1 1 0 0 0
3 3 2 3 2 1
```

Combinations /77

The defined function αCOMBω gives the index values for a character vector argument. We can now produce a character matrix of the possible combinations for two dogs being taken for a walk:

```
      'ABCD'[2 COMB 4]
CBBAAA
DDCDCB
```

When you define functions, you are simply writing expressions for the sequences of steps you want the APL computer to execute. The two sequences for generating combinations include two uses of the transpose function. These are necessary if you wish to eliminate blanks from the final character result. You might try to define functions to achieve the same goal *without* employing the transpose function. Such experiments should illustrate the "blank" aspect of the algorithm. You might also enjoy experimenting with definitions of your own that provide similar results in different ways. Such endeavors will add to your understanding as well as demonstrate the flexibility of APL.

Finally, let's compute all the combinations of three symbols chosen from a total of five:

```
       '⎕∇O*↑'[3 COMB 5]
O∇∇∇⎕⎕⎕⎕⎕⎕
**OO*OO∇∇∇
↑↑↑*↑↑*↑*O
```

EXERCISES

10.1 In how many ways can four breakfast items be selected from pancakes, sausage, eggs, juice and toast?

10.2 Generate a character matrix showing all of the combinations of the four breakfast items in problem 10.1.

10.3 If only one breakfast item is chosen in 10.1, provide all the combinations as a character array.

10.4 If five items are chosen for breakfast from 10.1, generate a character array showing all of the combinations.

10.5 Six commuters drive a Chevy, Honda, Ford, Lincoln, Mazda and Plymouth. Each driver is as likely as not to go through the Lincoln Tunnel on the way into the city. Provide a character array illustrating all of the combinations of four of the six cars that might pass through the tunnel on a given day.

10.6 What is the probability that four of the six cars in exercise 10.5 might go through the Lincoln Tunnel?

10.7 Suppose Matilda, Nancy, Ozzy, Penny, Quincy, Rosylyn and Seth are members of the Mathematics Club. Provide a character representation of all of the possible combinations of four of them being among the top four winners in the annual Mathematical Association of America Contest.

11/ PERMUTATIONS

$$!\omega$$
$$\vee \neq \omega$$
$$\alpha \vee \omega$$
$$PERM:(\alpha=+/\vee\neq(PO\alpha\rho\omega)\circ.=\iota\omega)/PO\alpha\rho\omega$$
$$NPERM:(\alpha!\omega)\times!\alpha$$
$$UO:\omega+\alpha\leq\omega$$
$$\alpha\uparrow\omega$$
$$PERMALT:UO\backslash PO\alpha\uparrow\phi 1+\iota\omega$$

In the final exercise in the preceding chapter, you were asked to calculate all the combinations of four out of seven club members. There are 35 different possible results. Let's take one and examine it in more detail.

Suppose the four winners are Nancy, Ozzy, Rosylyn and Seth. They are invited to an awards assembly, but they are not told until the awards are given who placed first, second, third and fourth. We want to know how many different "orders" in which the awards can be given. Each different ordering of the elements of a set is called a *permutation*. Another aspect of the problem is to provide the different orderings.

Finding the number of permutations is quite simple to calculate since first place could go to any of the four; and once the first award is given, there are three ways to give the next award. Two choices are possible for the third winner and the last person is automatically in fourth place. The product of these numbers provides the number of permutations:

```
      ×/4 3 2 1
24
```

The *factorial* function $!\omega$ provides another way to obtain the same result. That is, the factorial function gives the product of the natural numbers up to and including ω. We can use APL to visualize how factorial works:

```
      ι4
0 1 2 3
```

```
      1+ι4
1 2 3 4
```

We generate the counting integers in ω and add 1 to them because ⎕IO←0.

```
      ⌽1+ι4
4 3 2 1
```

```
      ×/⌽1+ι4
24
```

```
      !4
24
```

Since multiplication is commutative, the reverse function is unnecessary. However, it emphasizes the thinking process used in finding the number of permutations. That is, our problem involves four winners. As soon as we know who won first prize, there are three people eligible for second place. When the second winner is announced, only two people are candidates for third place, and so on.

Several strategies exist for generating permutations on an APL computer. A smaller version of the problem shows the technique. How might we award a first and second prize to three competitors, Andrea (0), Bart (1) and Chester (2), who enter a 100-meter race? The possible outcomes function yields the following result:

```
      PO 3 3
0 0 0 1 1 1 2 2 2
0 1 2 0 1 2 0 1 2
```

It is not possible for Andrea to win *both* first and second places, which is how the first column would be interpreted. The result we want is to have the elements in each column be unique orderings of two of the elements 0, 1 and 2. We need a technique for removing all columns which do not contain two unique winners.

When the first row of the result of PO 3 3 is compared with the 0 1 2 vector of possibilities using an equals outer product, the following matrix results:

```
      0 0 0 1 1 1 2 2 2∘.=0 1 2
1 0 0
1 0 0
1 0 0
0 1 0
0 1 0
0 1 0
0 0 1
0 0 1
0 0 1
```

The nine elements in the first column indicate which of the corresponding nine elements in 0 0 0 1 1 1 2 2 2 are zeros. The second column shows that the middle three elements are 1s. The final column shows that the last three elements are 2s.

When we use an equals outer product to compare the entire matrix generated by *PO* 3 3 with the vector of possibilities, the shape of the result is 2 9 3. The second row of the *PO* 3 3 matrix accounts for the second plane in the array:

```
      (PO 3 3)∘.=0 1 2
1 0 0
1 0 0
1 0 0
0 1 0
0 1 0
0 1 0
0 0 1
0 0 1
0 0 1

1 0 0
0 1 0
0 0 1
1 0 0
0 1 0
0 0 1
1 0 0
0 1 0
0 0 1
```

Permutations /83

The *logical or* function α∨ω generates a 1 if at least one of its arguments is a 1. Both α and ω must be Boolean. Applying an *or reduction* ∨≠ω to the first axis of the array with shape 2 9 3 gives a matrix whose shape is 9 3:

```
      ∨≠(PO 3 3)∘.=0 1 2
1 0 0
1 1 0
1 0 1
1 1 0
0 1 0
0 1 1
1 0 1
0 1 1
0 0 1
```

This array contains the information we need in order to select the columns containing two unique elements in the result of PO 3 3. The first row, 1 0 0, shows that only zeros are in the first column of the PO 3 3 matrix. The seventh row, 1 0 1, indicates the presence of a 0 and a 2 in the seventh column of the result of PO 3 3.

Since we are interested in counting the number of unique elements, we can add these Boolean numbers with a plus reduction.

```
      +/∨≠(PO 3 3)∘.=0 1 2
1 2 2 2 1 2 2 2 1
```

The positions of the 2s show which columns in the PO 3 3 matrix to choose:

```
      2=+/∨≠(PO 3 3)∘.=0 1 2
0 1 1 1 0 1 1 1 0
```

Replicating along the last axis selects the columns of the PO 3 3 matrix which are permutations of two distinct elements chosen from three different elements:

```
      (2=+/∨≠(PO 3 3)∘.=0 1 2)/PO 3 3
0 0 1 1 2 2
1 2 0 2 0 1
```

We can generalize these ideas with a *permutations* function αPERMω in which α represents the number of elements to be selected and ω is

84/ Permutations

the number of elements from which they are to be selected:

$PERM:(\alpha=+/\vee\neq(PO\alpha\rho\omega)\circ.=\iota\omega)/PO\alpha\rho\omega$

```
      2 PERM 3
0 0 1 1 2 2
1 2 0 2 0 1
```

We can use *PERM* to show permutations of letters or symbols rather than numbers:

```
     'ABC'[2 PERM 3]
AABBCC
BCACAB
```

The *PERM* function with arguments of 3 and 4 first generates the matrix result of *PO* 4 4 4. From the shape of the matrix, we know there are 64 columns or possible outcomes which must be searched for unique elements:

```
       ρPO 4 4 4
3 64
```

Once these have been located, the 24 columns provide the index positions of the character permutations:

```
     'ABCD'[3 PERM 4]
AAAAAABBBBBBCCCCCCDDDDDD
BBCCDDAACCDDAABBDDAABBCC
CDBDBCCDADACBDADABBCACAB
```

Suppose we are interested in finding all of the permutations of three distinct letters chosen from the word *PEONY*. In these permutations, *ONE* and *EON* are considered to be different. There are 3!5 (10) combinations and !3 (6) permutations of each one of the combinations. Their product, 60, is the number of permutations of 3 items chosen from 5 elements. The defined function for calculating the *number of permutations*, $\alpha NPERM\omega$, gives the number of ways α items can be selected from ω elements:

$NPERM:(\alpha!\omega)\times!\alpha$

Permutations

```
      3 NPERM 5
60
```

This is the same as:

```
      x/5 4 3
60
```

Note that when α and ω are equivalent, the result is the same as !ω:

```
      5 NPERM 5
120
```

This is true since 5!5 is 1.

Suppose we want 2 rather than 3 of the 5 symbols:

```
      2 NPERM 5
20
```

```
      x/5 4
20
```

Our original problem of finding the unique orderings of four elements chosen from seven requires searching the 2401 columns of the matrix created by PO 7 7 7 7 for those with four distinct elements. Since 4!7 or 35 combinations of 4 different students might be the winners of the Mathematical Association of America Contest and there are !4 or 24 different permutations of the selected students, the number of permutations of 4 students chosen from 7 club members is:

```
      24×35
840
```

This is also the product of the first four elements of !7.

```
      x/7 6 5 4
840
```

That is, of the 2401 columns in the possible outcomes matrix, 840 will be selected. Some small APL systems do not have enough workspace to generate all the possible outcomes needed in order to find the permutations with this strategy.

86/ Permutations

On systems that allow defined functions to be used with the scan operator, there is a strategy for generating the permutations that does not require building a very large array and then eliminating from it those outcomes which don't suit our criteria. It provides a very different and fascinating way of generating permutations. To understand what follows, you need to know how *scan* and *reduction* work. We begin with a discussion of reduction, because scan is defined in terms of reduction.

Reduction of a vector by a function (for example, addition) is defined as being equivalent to inserting the addition function between each pair of elements. Thus, both of these APL expressions give the same result:

```
      +/1 2 3 4
10
```

```
      1+2+3+4
10
```

If there is only one element in the vector, we can't put the symbol "between" anything. By definition, reduction of a one-element vector is equal to that element. Thus:

```
      +/3
3
```

The scan of a vector by a function (for example, multiplication) is defined as being equivalent to *reducing* successively longer prefixes of the vector by the function. Therefore,

```
      ×\1 2 3 4
1 2 6 24
```

is the same as

```
      (×/1),(×/1 2),(×/1 2 3),×/1 2 3 4
1 2 6 24
```

To investigate how we can use the scan operator to find permutations, let's again consider awarding first and second prizes for our three competitors, Andrea, Bart and Chester. Bear in mind that we're seeking the same result generated by α*PERM*ω:

```
      2 PERM 3
0 0 1 1 2 2
1 2 0 2 0 1
```

Suppose we begin with *PO* 3 2:

```
      PO 3 2
0 0 1 1 2 2
0 1 0 1 0 1
```

This is an appropriate start in solving a permutations problem because the matrix result of *PO* has the correct number of columns from the beginning. There are three choices for the first element in each column of the result and two for the second element. We want to insure that the elements in each column are unique. As we found when representing possible outcomes as characters, it is necessary to correct the index values in all but the first row of the result *PO*ω.

Letting α be the first row and ω the second, we can compare them with the less than or equal function:

```
      0 0 1 1 2 2≤0 1 0 1 0 1
1 1 0 1 0 0
```

If we add the result of α≤ω to ω, we can obtain the second row in the matrix result of α*PERM*ω. That is, we need to add 1 to each element in the second row of the *PO* 3 2 matrix whenever it is greater than or equal to the element above it in the first row. Putting it another way, whenever the value in the first row is less than or equal to its corresponding value in the second row, we must add 1 to the latter.

```
      0 1 0 1 0 1+1 1 0 1 0 0
1 2 0 2 0 1
```

By defining this process in a *unique orderings* function α*UO*ω, we can use it to scan the rows of the matrix generated by *PO* 3 2:

$UO:\omega+\alpha\leq\omega$

```
      UO⍀PO 3 2
0 0 1 1 2 2
1 2 0 2 0 1
```

This result agrees with the permutation matrix created by

88/ Permutations

2 *PERM* 3. (If your system does not permit the scan operator to take a defined function as its left argument, you will get an error message when you try to execute $UO\backslash PO\ \omega$.)

Since the scan operator is being applied to each column in the *PO* 3 2 matrix, the first element in each column remains unchanged. The 2 in the second column is the result of $\omega+\alpha\leq\omega$ being applied where α is 0 and ω is 1:

```
      1+0≤1
2
```

In this fashion, each element in the second row is generated.

Suppose we want the APL computer to generate all of the possible orders for three elements taken three at a time. Put in terms of our 100-meter race problem, we want all the unique ways Andrea, Bart and Chester can place. Once we've established the winners of first and second place, it's obvious that the remaining person came in third.

```
      PO 3 2 1
0 0 1 1 2 2
0 1 0 1 0 1
0 0 0 0 0 0
```

Scanning the columns of the *PO* 3 2 1 matrix with the function $\alpha UO\omega$ creates the same first and second rows we found earlier. But, the third row requires a longer sequence of steps. First, the second and third rows are processed by $\omega+\alpha\leq\omega$:

```
      0 0 0 0 0 0+0 1 0 1 0 1≤0 0 0 0 0 0
1 0 1 0 1 0
```

Then the 1 0 1 0 1 0 result becomes ω, and the first row of the *PO* 3 2 1 matrix is α:

```
      1 0 1 0 1 0+0 0 1 1 2 2≤1 0 1 0 1 0
2 1 2 0 1 0
```

```
      UO\PO 3 2 1
0 0 1 1 2 2
1 2 0 2 0 1
2 1 2 0 1 0
```

This method has the advantage of generating only the desired number

Permutations /89

of columns and then manipulating them to provide the appropriate permutations.

Suppose we were finding all the permutations of seven elements taken seven at a time. One of the columns of the matrix created by PO 7 6 5 4 3 2 1 is 4 4 2 2 0 1 0. It is very useful to see how $\omega+\alpha\leq\omega$ with the scan operator takes this column and creates a unique permutation of the elements 0 1 2 3 4 5 6. The first element is 4. The second is $4+4\leq 4$ or 5. The third element is found in two steps:

$$2+4\leq 2$$
2

$$2+4\leq 2$$
2

The fourth element is derived by:

$$2+2\leq 2$$
3

$$3+4\leq 3$$
3

$$3+4\leq 3$$
3

The fifth element of the PO column remains a 0:

$$0+2\leq 0$$
0

$$0+2\leq 0$$
0

$$0+4\leq 0$$
0

$$0+4\leq 0$$
0

Somewhat more instructive is how the sixth element (which is 1) is transformed into a 6 in the result of the UO scan:

	1+0≤1
2	
	2+2≤2
3	
	3+2≤3
4	
	4+4≤4
5	
	5+4≤5
6	

What is happening here is that the element we're working with is compared with the preceding element and it is augmented by one if it is not the smaller element. This new element, in turn, is compared with the element to its left, and so on. This procedure removes all duplications. The *PO* column's final element, a 0, is increased by 1 once to become a 1 in the result of the *UO* scan:

	0+1≤0
0	
	0+0≤0
1	
	1+2≤1
1	
	1+2≤1
1	
	1+4≤1
1	
	1+4≤1
1	

Taking these results all together we have:

```
      UO\4 4 2 2 0 1 0
4 5 2 3 0 6 1
```

Notice how, beginning with a vector in which there are repeated elements, we end up with a vector in which all the elements are distinct—we have developed a permutation vector.

We can now define an alternate permutations function α*PERMALT*ω to generate all the unique orderings of α items selected from ω elements:

PERMALT:*UO**PO*α↑ɸ1+ιω

The *take* function α↑ω selects elements from an array. The left argument specifies the number of elements to be taken from the array in the right argument:

```
      ɸ1+ι5
5 4 3 2 1

      2↑ɸ1+ι5
5 4

      2 PERMALT 5
0 0 0 0 1 1 1 1 2 2 2 2 3 3 3 3 4 4 4 4
1 2 3 4 0 2 3 4 0 1 3 4 0 1 2 4 0 1 2 3
```

Inspection of this matrix reveals !2 times 2!5, or 20 different combinations. That is, 20 *pairs* of elements appear in the result. Indexing the vector '□∇○⋆↑' produces the permutations as a character matrix:

```
      '□∇○⋆↑'[2 PERMALT 5]
□□□□∇∇∇∇○○○○⋆⋆⋆⋆↑↑↑↑
∇○⋆↑□○⋆↑□∇⋆↑□∇○↑□∇○⋆
```

In a similar fashion 3 *PERMALT* 4 begins by creating *PO* 4 3 2. The scan generates indexes for a character matrix identical to the one we found earlier:

```
      'ABCD'[3 PERMALT 4]
AAAAAABBBBBBCCCCCCDDDDDD
BBCCDDAACCDDAABBDDAABBCC
CDBDBCCDADACBDADABBCACAB
```

The letters in each column are three different letters chosen from *ABCD*.

In summary, α*NPERM*ω and α!ω provide the number of permutations and combinations respectively. In a similar fashion, α*PERM*ω (or α*PERMALT*ω) and α*COMB*ω generate numeric arrays of α items selected from ω elements with and without order as a criterion.

EXERCISES

11.1 Find all the different permutations of the word *CORN*. Use a transpose ⍉ so that the permutations are the rows of the matrix. To understand the amount of computation required to solve this type of problem (and to test the capacity of your machine), find all of the different permutations of the word *ACORN*.

11.2 How many different permutations of three-letter words can be chosen from *ACORN*?

11.3 Find all the three-letter words which can be chosen from *ACORN*.

11.4 You have four flags that are red, blue, yellow and green. Generate a character matrix of all of the different possible arrangements on a flagpole using all four flags.

11.5 If you add an orange and a purple flag, how many 4-flag signals can you send?

11.6 How many permutations are there of 5 cards chosen from a standard bridge deck of 52 cards?

11.7 In how many ways can three cars be put in five garages?

11.8 How many words of all possible sizes can you make from the characters in *SOLAR*? Assume no repetition of letters.

11.9 When you find all the permutations of four letters from the word *EDUCATION* with *PERM* and *PERMALT*, how many possible out-

Permutations /93

comes does each function generate in the process of returning its result? That is, what is the number of numbers generated by *PERM* compared to the number of numbers generated by *PERMALT*?

12/ GEOMETRIC DISTRIBUTIONS

$$GP: \alpha \times (1-\alpha) \star \iota\omega$$
$$\alpha \triledown \omega$$
$$TA0: 0\ 0\ 0\ 4\triledown(\iota\rho\omega),[.5]\omega$$
$$TA1: 0\ 0\ 0\ 4\triledown(1+\iota\rho\omega),[.5]\omega$$
$$BD: TA0\alpha BP\omega$$
$$GD: TA1\alpha GP\omega$$
$$\lfloor\omega$$
$$\alpha \leq \omega$$
$$\alpha \wedge \omega$$
$$\wedge \backslash \omega$$

In the chapter on the binomial distribution, we considered a basketball player whose probability of success on an individual free throw was 25 percent. Here's the table of probabilities for that player making 0, 1, 2 or 3 baskets out of 3 free throws:

```
        (ιρ3 BP .25),[.5]3 BP .25
0          0.4219
1          0.4219
2          0.1406
3          0.01563
```

What if the player begins taking shots at the basket and continues until a basket is made? The probability of success on the first throw is .25. However, making a basket on the second throw occurs only if there is a miss on the first throw. The probability is .25×.75. Two misses and a basket are necessary for success on the third try; that is, .25×.75×.75. To generate the probability distribution of the first 7 trials, we begin with the misses and raise .75 to each of the powers 0, 1, 2, 3, 4, 5 and 6:

```
      .75*0 1 2 3 4 5 6
1 0.75 0.5625 0.4219 0.3164 0.2373 0.178
```

Geometric Distributions /95

Multiplying by .25 contributes the one final success to each possible outcome:

```
      .25×.75*0 1 2 3 4 5 6
0.25 0.1875 0.1406 0.1055 0.0791 0.05933 0.04449
```

Thus, the fifth element in the result is the probability of four misses and a basket on the fifth throw.

We can write an APL expression for the first success and use it to define a *geometric probabilities* function $\alpha GP \omega$. The argument α gives the probability of success. Failure is $1-\alpha$ and ω is the number of trials. (The name "geometric" comes from the traditional differentiation between arithmetic and geometric progressions. A geometric series is a sequence of terms each of which is a constant multiple of the immediately preceding term.)

```
GP:α×(1-α)*ιω
```

When we form a table of the geometric probability distribution, we begin counting with 1 rather than 0, as we did with the binomial disribution.

```
        (1+ιρ.25 GP 7),[.5].25 GP 7
1       0.25
2       0.1875
3       0.1406
4       0.1055
5       0.0791
6       0.05933
7       0.04449
```

There is no guarantee of success, so theoretically the distribution is infinite. However, the probability of a first success diminishes as successive terms in the sequence are found. The conditions necessary for geometric experiments are that the trials must have a constant probability of success and that trials are continued until a success is obtained.

We can generate more attractive and readable results with the *character representation* function $\alpha \bar{\top} \omega$. The vector in α has a pair of elements for each column of the array in ω. The first element of each pair indicates the number of characters allowed for the columns. The second element of the pair indicates the number of decimal places. When the first element of the pair is zero, the system determines the number of columns.

```
      0 0 0 4⍕(1+⍳⍴.25 GP 7),[.5].25 GP 7
1 0.2500
2 0.1875
3 0.1406
4 0.1055
5 0.0791
6 0.0593
7 0.0445
```

The second element of the second pair in the left argument of ⍕ specifies that the numbers in the second column of the result are to be displayed to 4 decimal places (ten-thousandths).

Two defined functions for generating *tables*, $TA0\omega$ and $TA1\omega$, include 0 and 1 in their names to indicate the origin of the counting numbers in the first column of their matrix results:

$TA0:0\ 0\ 0\ 4⍕(\iota\rho\omega),[.5]\omega$
$TA1:0\ 0\ 0\ 4⍕(1+\iota\rho\omega),[.5]\omega$

The *binomial distribution of probabilities* function $\alpha BD\omega$ generates a table of the results of $\alpha BP\omega$.

$BD:TA0\alpha BP\omega$

A distribution of 3 trials by the player with a 25 percent success probability for each free throw is:

```
      3 BD .25
0 0.4219
1 0.4219
2 0.1406
3 0.0156
```

$\alpha GD\omega$ is a defined function which shows a *geometric distribution of probabilities*. That is, $\alpha GD\omega$ displays the table of results created by $\alpha GP\omega$.

$GD:TA1\alpha GP\omega$

The likelihood of our player with a 25 percent probability of success making the first basket on each of 7 throws is:

Geometric Distributions

```
        .25 GD 7
1 0.2500
2 0.1875
3 0.1406
4 0.1055
5 0.0791
6 0.0593
7 0.0445
```

The probability of the first basket on the fifth throw is 0.0791.

Suppose we want to simulate the same basketball player making 100 attempts of 3 throws each. We can generate such a simulation with the following expression:

```
        FT+/1>?3 100ρ4
0 41
1 48
2  9
3  2
```

?3 100ρ4 creates a 3-row by 100-column matrix of random integers from 0 to 3 inclusive.

```
        1>0 1 2 3
1 0 0 0
```

When 1 is compared with each of the equally possible elements generated by the roll function, the result 1 0 0 0 has one 1 and three 0s. Thus, if we consider 1 to represent a success, we can use the greater than function to establish the probability of success for our basketball player. Putting it another way, one is always greater than each possible integer from zero to three inclusive 25 percent of the time. Once success and failure are represented as a Boolean matrix, we can tally the sums of each column with our frequency table function. The second column of the result in the preceding example contains a "simulation vector" with the values 41 48 9 2. We can compare this with the expected distribution of 100 samples:

```
        100×3 BP .25
42.19 42.19 14.06 1.563
```

The *floor* function ⌊ω facilitates the comparison. Floor rounds down so that each element in its result is the largest integer in the corre-

98/ **Geometric Distributions**

sponding element of its argument. For example,

```
      ⌊22.2 6.16 5.5
22 6 5
```

```
      ⌊100×3 BP .25
42 42 14 1
```

We can use the floor function to round up by adding .5 to each element of ω before we apply it:

```
      ⌊.5+100×3 BP .25
42 42 14 2
```

The study of statistics includes the analysis of a simulation vector such as 41 48 9 2 and its expected distribution 42 42 14 2.

Let's consider our player making the first basket in seven trials. First, we generate a sample of seven random elements:

```
      ?7ρ4
3 2 1 3 0 1 0
```

We can use the *less than or equal* function α≤ω to examine the result of the roll function.

```
      1≤3 2 1 3 0 1 0
1 1 1 1 0 1 0
```

Notice that using the less than or equal function implies that 0 represents a success:

```
      1≤0 1 2 3
0 1 1 1
```

The probability of our player's success is still .25 because the less than or equal relationship is true 25 percent of the time. Letting zero stand for a success is useful because it is easy to find the first 0 in a vector with APL. The *logical and* function α∧ω requires that both arguments be Boolean. An element of its result is a 1 only if the corresponding elements in both arguments are 1:

```
      0 0 1 1∧0 1 0 1
0 0 0 1
```

When an *and scan* ∧\ω is applied to the vector result of ?7ρ4 in our example, we can see a convenient ordering which we can exploit in solving our problem:

```
      ∧\1 1 1 1 0 1 0
  1 1 1 1 0 0 0
```

Reading from left to right, the first zero in the result corresponds to the first zero in the argument, *and the rest of elements in the result are also zero.*

If we take the sum of the result of the and scan of the less than or equal relationship, we'll compute the position of the last failure. That is, since we're now considering zero to represent a success, the sum of the 1s shows us where the failures end.

```
      +/∧\1≤3 2 1 3 0 1 0
  4
```

To find the position of the first success, we need only add 1:

```
      1++/∧\1≤3 2 1 3 0 1 0
  5
```

The following expression simulates 100 samples of 7 free throws by our basketball player:

```
      FT 1++⌿∧⍀1≤?7 100ρ4
  1 29
  2 14
  3 13
  4  6
  5 12
  6  9
  7  6
  8 11
```

The and scan and the plus reduction apply to the columns of the random matrix. Note that the 11 cases of a basket on the eighth free throw simply indicate that a success has not occurred in the first 7 attempts.

```
         ⌊.5+100×.25 GP 7
  25 19 14 11 8 6 4
```

100/ **Geometric Distributions**

The vector 25 19 14 11 8 6 4 comes from the theoretical geometric distribution of success in obtaining the first basket in any one of 7 successive free throws. It accounts for only 87 of the 100 events. Approximately 13 percent of the time, a basket might not have been made by the eighth throw.

EXERCISES

12.1 If 5 percent of the customers who enter The Tape Deck buy a tape, what is the theoretical probability distribution for 0, 1, 2, 3 and 4 sales in the next hour, if 4 customers enter the shop during that time?

12.2 Simulate 50 samples of the 4 customers described in 12.1.

12.3 If 5 percent of the customers who enter The Tape Deck buy a tape, what is the theoretical probability distribution of the likelihood for the first sale to occur among the first 10 customers?

12.4 Simulate 200 samples of the 10 customers described in 12.3.

13/ THE DEAL FUNCTION

$$DBD:TA0((\iota 1+\omega)!1\uparrow\alpha)\times((\phi\iota 1+\omega)!|-/\alpha)\div\omega!1\downarrow\alpha$$
$$-/\omega$$
$$|\omega$$
$$\alpha?\omega$$
$$\pm\omega$$

To explore a variation of the binomial probability distribution, we can consider a bowl of 40 assorted jelly beans. Assuming that we prefer licorice jelly beans, we'll look at some aspects of choosing this flavor. When the bowl contains 10 of them, the probability of our reaching in and getting a licorice candy is .25. However, the probability of our getting two licorice jelly beans is not the same as the likelihood of our basketball player getting two baskets in two free throws (.25*2 or 0.0625). This is because the two events are not independent of one another.

Two events are dependent if the occurrence of one event is influenced by an occurrence of the other. This is the first example we have encountered in which the trials are dependent. The number of combinations of 2 of the 10 licorice jelly beans is 2!10 or 45. Since 2 candies can be selected in 2!40 or 780 ways, the probability of our getting 2 licorice jelly beans is:

```
      (2!10)÷2!40
0.05769
```

If we select three jelly beans, the probability of getting two licorice ones is the product of the number of ways we can get two licorice and the number of ways we can get one of another flavor:

```
      2!10
45

      1!30
30

      (2!10)×1!30
1350
```

The number of possible combinations of candies when 3 out of 40 jelly beans are selected is:

```
        3!40
9880
```

The probability of getting exactly two licorice in a handful of three is:

```
        (2!10)×(1!30)÷3!40
0.1366
```

Calculating the distribution of the probabilities for selecting 0, 1, 2 or 3 licorice jelly beans in handfuls of 3 can be developed as follows. First, create a vector of the possible combinations for getting 0, 1, 2 or 3 licorice beans:

```
        0 1 2 3!10
1 10 45 120
```

Then, generate a vector representing all the combinations for the other flavors. Notice that getting no licorice beans implies getting 3 of the others. Similarly, the presence of 1 licorice bean means the other two are different flavors, and so on.

```
        3 2 1 0!30
4060 435 30 1
```

Compute the number of ways 3 candies can be chosen from 40:

```
        3!40
9880
```

The product of the vectors divided by the number of possible outcomes generates the probabilities that 0, 1, 2 or 3 are licorice. The final result is the desired dependent binomial distribution.

```
        (0 1 2 3!10)×(3 2 1 0!30)÷3!40
0.4109 0.4403 0.1366 0.01215
```

We can generalize the solution to our jelly bean problem and define $\alpha DBD\omega$ to provide the *dependent binomial distribution*.

$DBD:T\text{AO}((\iota 1+\omega)!1\uparrow\alpha)\times((\phi\iota 1+\omega)!|-/\alpha)\div\omega!1\downarrow\alpha$

The 2-element vector α contains the number of desired items and the number of all the elements. For our problem, α is 10 40. The right argument is the number of items to be selected. Since we've been considering handfuls of 3, that is the value of ω in this problem.

```
      3!1↓10 40
9880
```

We use a *minus reduction* denoted by the symbols -/ with the *absolute value* function |ω to generate the number of "other" or "not desired" items. The absolute value function is also called the *magnitude* function. It takes any array as its argument. Its result is identical to its argument except that any negative elements are made positive.

```
      -/10 40
¯30

      |-/10 40
30

      (⌽⍳1+3)!|-/10 40
4060 435 30 1

      (⍳1+3)!1↑10 40
1 10 45 120

      10 40 DBD 3
0  0.4109
1  0.4403
2  0.1366
3  0.0121
```

Because the events are not independent, the dependent binomial distribution differs from the binomial distribution of probabilities for 0, 1, 2 or 3 free throws for the basketball player with a 25 percent probability of success for each throw.

```
      3 BD .25
0  0.4219
1  0.4219
2  0.1406
3  0.0156
```

The APL computer generates dependent random numbers as well

The Deal Function /105

as independent random numbers. The *deal* function α?ω creates a vector of α elements selected from ιω without repetition of elements. This is exactly what is needed to simulate dependent events.

```
      ι10
0 1 2 3 4 5 6 7 8 9

      3?10
4 0 7

      3?10
6 5 8
```

The lack of repetition in these examples differs from the result of the roll function:

```
      3ρ10
10 10 10

      ?3ρ10
8 5 8
```

When deal is used, there are no repeated elements in the result.

To select three jelly beans we use:

```
      3?40
25 5 30
```

When 10 is compared with ι40 using the greater than function, the result is a vector with ten 1s and thirty 0s. Thus, 10>40?40 gives a vector with a random distribution of ten 1s among thirty 0s. When we compare 10 with 25 5 30 the 1s in the result are the licorice jelly beans:

```
      10>25 5 30
0 1 0
```

10 is greater than 25 5 30 once, so our "handful" has one licorice jelly bean:

```
      +/10>25 5 30
1
```

If we want 5 simulations we must use the deal function 5 times:

```
      (3?40),(3?40),(3?40),(3?40),(3?40)
5 0 17 22 3 21 34 20 35 0 17 11 37 20 33
```

Although there are repetitions in the 15-element vector, there are no repetitions in each of the 5 sets of 3 elements. We shorten the typing of the cumbersome expression by using reshape to create a character vector:

```
      34ρ'(3?40),'
(3?40),(3?40),(3?40),(3?40),(3?40)
```

The character vector '(3?40),' has 7 elements. Repeating it 5 times requires 35 characters. Using 34 as the left argument of reshape eliminates the final comma.

The *execute* function ⍎ω evaluates and executes the character scalar or vector in ω. Thus, ω must contain a valid APL expression. The effect of the execute function in the following example is twofold: 1) the APL computer generates 5 deals; 2) the elements in the result are numeric data on which calculations can be made.

```
      ⍎34ρ'(3?40),'
10 22 24 37 28 9 7 2 1 32 29 0 5 25 17
```

We can generate 5 different deals in matrix form using reshape:

```
      5 3ρ⍎34ρ'(3?40),'
14 24 18
 5 17 29
13  4 26
28 15 22
36  8 35
```

Note that the "handfuls" are in the rows of the simulation matrix. Here's a frequency table of 100 samples of handfuls of 3 jelly beans:

```
      FT+/10>100 3ρ⍎699ρ'(3?40),'
0 35
1 44
2 19
3  2
```

The Deal Function /107

Each of the 100 trials are represented in the 100 rows of the matrix.
Another simulation of 100 samples of 3 free throws is shown for comparison.

```
      FT+/1>?3 100ρ4
0 44
1 39
2 16
3  1
```

The major difference in the two simulations is in the use of the roll or the deal function to generate independent or dependent events respectively. In both cases, the events are mutually exclusive. Since all of the probabilities are assigned to a finite number of results in each case, the results for both are also called exhaustive. The sum of the probabilities is 1.

EXERCISES

13.1 Suppose 20 out of 50 students listen to a local rock station while doing their homework at home one evening. Ten students are polled and asked if they had listened to the station on the previous evening. Generate the theoretical probability distribution of 0 through 10 students saying yes to the question.

13.2 Simulate 40 samples of the 10 students in 13.1 being polled.

13.3 From past experience, it is known that only 40 percent of the students asked to spell "accommodate" in spelling contests are successful. If each of the ten contestants in the first round of a spelling bee are asked to write the word on a piece of paper, find the probability distribution for the number of students likely to be successful and advance to the second round.

13.4 Simulate 40 samples of 10 students participating in a spelling contest like the one described in 13.3.

13.5 Over a long period of time at the beach in the summer, you observe that 1 seagull in 13 catches a fish when diving for them. Find the frequency distribution for 0 through 4 fish being caught when samples of 4 seagulls are considered.

13.6 Simulate 100 samples of 4 seagulls fishing under the conditions described in 13.5.

13.7 At the beach in the summer, you play many hands of bridge. Find the theoretical distribution for the number of aces in each hand.

13.8 Simulate sampling 100 bridge hands and find the frequency distribution of aces.

13.9 "Shuffle" the deck of cards you created in 6.5.

14/ THE WORLD SERIES PROBLEM

How many different permutations of the letters in the word *SPINACH* are possible? We found earlier that 7 different elements can be arranged in !7 different ways:

!7
5040

What if the word has two elements which are the same, as in *SPANISH*? Once an arrangement such as *SANSHIP* is found, the two *S*s can be interchanged. Thus, there are only half as many possibilities as there were for *SPINACH*.

(!7)÷!2
2520

÷/!7 2
2520

Suppose the word is *SUCCESS*. We can obtain a combination like *ECSSUSC*, but it is necessary to divide the result of !7 by the product of !2 and !3 to take into account the interchanging of the *C*s and the *S*s respectively:

(!7)÷×/!2 3
420

The number of arrangements of the letters in *MISSISSIPPI* is:

(!11)÷×/!4 4 2
34650

A very interesting variation of these problems involves sequences which have only two distinct elements such as *ENEENNE*:

(!7)÷×/!4 3
35

The number of combinations obtained by 4!7 and 3!7 is also 35:

```
      4!7
35
```

```
      3!7
35
```

If we wish to travel 3 blocks North and 4 blocks East to go from point *A* to point *B*, one path might be represented by *NEEENEN*. We have found that there are 3!7 or 4!7 or 35 different paths from *A* to *B*. Figure 5 illustrates this *concept* by showing just one of the possible routes between a different set of points, *a* and *b*.

A somewhat different perspective can be gained from another problem. Assume that we want to separate 10 identical cars into 3 groups, with at least 1 car in each group. We can create a character vector in which the symbol ☐ represents a car and ∧ represents the possible positions separating the "cars" into 3 groups:

```
    19ρ'☐∧'
☐∧☐∧☐∧☐∧☐∧☐∧☐∧☐∧☐∧☐
```

Selecting 2 of the 9 positions marked by ∧ separates the 10 items in 3 parts:

```
      2!9
36
```

The third and eighth of the nine possible positions for the two ∧ symbols separate the cars into one of the 36 different possible sequences. Its character representation looks like this:

☐☐☐∧☐☐☐☐☐∧☐☐

Using the idea of permutations of 2 different kinds of elements, we can determine the likelihood of 2 equally matched teams completing the World Series in 4, 5, 6 or 7 games.

If we assume the American League takes the Series, the final game must be an American League win. Letting *A* stand for an American League win, a character sequence prior to the final game must contain three *A*s such as *AAA*, *ANAA*, *ANNAA* or *NNNAAA*, depending upon whether the Series ends in 4, 5, 6 or 7 games. Since there are only two options, the number of permutations are:

112/ The World Series Problem

Fig. 5

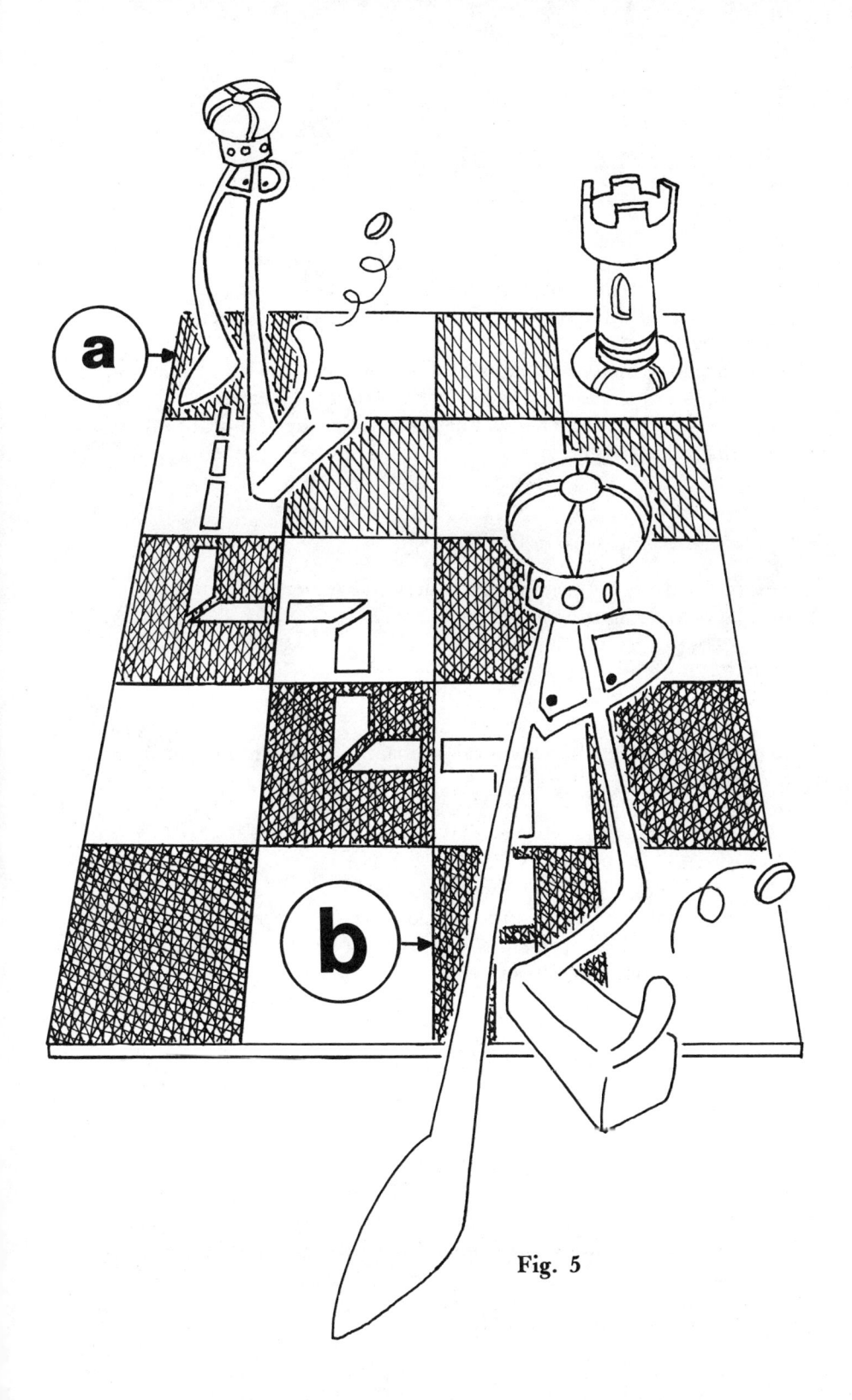

```
      3!3 4 5 6
1 4 10 20
```

```
      +/1 4 10 20
35
```

There are 10 different sequences in which the American League wins in 5 games and 35 different ways in which the American League wins the World Series.

Sequences in which the National League wins end in *N* and the games in the 4 sequences must have 3 prior National League wins such as *NNN*, *NANN*, *NANAN* or *ANNAAN*. The possibilities are identical to those for the American League:

```
      3!3 4 5 6
1 4 10 20
```

There are 70 different possible sequences altogether. The probability of the National League winning in 5 games is:

```
      10÷70
0.1429
```

Each of the 2 leagues has identical probabilities of winning in 4, 5, 6 or 7 games:

```
      1 4 10 20÷70
0.01429 0.05714 0.1429 0.2857
```

Designing a simulation of 70 "World Series" competitions is now quite easy. Letting a 1 represent a win by the National League, a plus scan shows the accumulation of wins:

```
      ?7ρ2
1 1 0 1 1 0 1
```

```
      +\1 1 0 1 1 0 1
1 2 2 3 4 4 5
```

The National League wins the Series when the first 4 occurs in the vector. When 4 is compared with 1 2 2 3 4 4 5 using the greater than function, a 1 indicates all games before the winning game:

```
        4>1 2 2 3 4 4 5
1 1 1 1 0 0 0
```

Adding 1 to the plus reduction of the vector provides the information we want. The National League wins in the fifth game of the series:

```
        1++/4>+\1 1 0 1 1 0 1
5
```

When the American League wins in the sixth—or in any game—the result is an 8. (Test this for yourself.) Thus, an 8 is interpreted as an American League win:

```
        1++/4>+\0 1 0 0 1 0 1
8
```

The 70 simulations can be generated in the columns of a matrix and displayed in a frequency table:

```
        FT 1++/4>+\?7 70p2
4  5
5  7
6  12
7  12
8  34
```

In a simulation of 70 games, the National League wins 36 times in 4, 5, 6 and 7 games with frequencies of 5, 7, 12 and 12 respectively. In 34 instances of the 70 simulations, the American League wins.

EXERCISES

14.1 Eleven points lie on the circumference of a circle. How many inscribed convex hexagons can be drawn using the points as vertices?

14.2 How many different permutations of the symbols □□□□□△△△ooooo are possible?

14.3 A class of 10 girls and 8 boys form a line with the girls arranged in ascending order of height and the boys similarly. In how many ways can this be done? (No two students have the same exact height.)

14.4 Two friends can't decide how to spend an evening together since the first wants to go to an ice hockey game and the second wants to go to a movie. They agree to flip a coin until one gets five successes. The first person agrees to take "heads" or 1s. Find the probability distribution for going to an ice hockey game as determined by 5, 6, 7, 8 and 9 flips of the coin.

14.5 Simulate 60 trials of the situation described in 14.4.

15/ THE FIRST ACE

$$DGD:TA1((^{-}1+1\uparrow\alpha)!\omega\uparrow\phi\iota1\downarrow\alpha)\div!/\alpha$$

A classic problem in probability theory is to determine the probability of turning up the first ace from a shuffled deck of cards. While the geometric distribution function provides the probabilities of the first occurrence of independent events, drawing the first ace involves dependent ones.

Let's look at a similar problem. Suppose we have 10 licorice jelly beans in a jar containing a total of 40. If we dip into the bowl 7 times and select 1 candy each time, what is the probability that we'll get a licorice bean on any of the 7 attempts? Generating the probability distribution which answers this question involves the product of probabilities, but these change each time we remove another bean from the jar.

Letting L stand for licorice and A for "assorted," we can represent getting a licorice bean on the first try by the one-element character vector L. AL AAL $AAAL$ $AAAAL$ $AAAAAL$ $AAAAAAL$ illustrate the pattern of flavors we'd see if the first licorice bean was obtained on the second through seventh draw, respectively. The only variability is in the remaining 39, 38, 37, 36, 35, 34, 33 orders of A and L.

The number of possibilities for all of the arrangements of the 40 jelly beans is:

```
      10!40
8.477E8
```

The presence of $E8$ in the result indicates that the decimal belongs 8 places to the right of the position shown, so that $8.477E8$ is 847,700,000. (Recall that we assigned a value of 4 to the system variable $\square PP$. The actual result is 847,660,528.) This is the number of permutations of the 10 licorice and 30 assorted flavor jelly beans. Drawing a licorice bean on the fourth try implies that 36 beans are left in the bowl and 9 of them are licorice. Thus, there are $9!36$ or $9.414E7$ arrangements for the remaining 9 licorice beans in a total of 36 jelly beans. Dividing those permutations by all possible arrangements of the 40 jelly beans gives the probability of drawing the first licorice bean on the fourth try:

```
      (9!36)÷10!40
0.1111
```

To calculate the probabilities for drawing the first licorice bean on each of the first 7 attempts, we need the result of

```
9!39 38 37 36 35 34 33
```

because 9 licorice jelly beans always remain, and the number of assorted jelly beans diminishes with each successive pick:

```
        9!39    38     37    36     35     34     33
2.119E8 1.63E8 1.244E8 9.414E7 7.061E7 5.245E7 3.857E7
```

Division produces the desired probabilities:

```
     (9!39 38 37 36 35 34 33)÷10!40
0.25 0.1923 0.1468 0.1111 0.0833 0.06188 0.0455
```

We can define a *dependent geometric distribution* function $\alpha DGD\omega$ in which α is the pair of numbers of successful and total outcomes respectively, and ω is the number of trials.

$DGD:TA1((^{-}1+1\uparrow\alpha)!\omega\uparrow\phi 1 1\downarrow\alpha)\div !/\alpha$

```
      10 40 DGD 7
1 0.2500
2 0.1923
3 0.1468
4 0.1111
5 0.0833
6 0.0619
7 0.0455
```

Calculating the distribution involves three major components:

```
     !/10 40
8.477E8

       7↑ψι40
39 38 37 36 35 34 33
```

118/ The First Ace

```
        ¯1+10
9
```

Together they generate the desired result:

```
    (9!39 38 37 36 35 34 33)÷8.477E8
0.25 0.1923 0.1468 0.1111 0.08329 0.06187 0.0455
```

The dependent geometric distribution differs from the geometric distribution because the events it describes are dependent. Compare the probabilities for obtaining the first licorice jelly bean in 7 draws with a basketball player's chances of shooting the first basket on each of the first seven attempts:

```
      .25 GD 7
1 0.2500
2 0.1875
3 0.1406
4 0.1055
5 0.0791
6 0.0593
7 0.0445
```

The dependent geometric distribution also differs from the dependent binomial distribution. The dependent geometric distribution shows the probabilities of obtaining the first occurrence of an event. The dependent binomial distribution gives the probabilities of obtaining a specific number of successes in a given trial.

In a simulation to find the first occurrence of a licorice jelly bean in 100 samples of 7 draws, the and scan applies to the rows of a matrix which contain 7 successive deals of the first 40 integers:

```
      FT 1++/∧\10≤100 7ρ⍳699ρ'(7?40),'
1 26
2 22
3 12
4 11
5  2
6 13
7  1
8 13
```

The First Ace /119

A success on the eighth draw is interpreted as "no success" in the first seven.

Compare this with a simulation of the first basket obtained by a basketball player in 100 samples of 7 free throws:

```
      FT 1++≠∧⍳1≤?7 100ρ4
1 27
2 23
3 10
4  8
5  8
6  4
7  5
8 15
```

The results in the two preceding examples are very similar. However, in one case the simulation depends upon the deal function where repetition is not allowed and the other is based upon the roll function which permits repetition. Simulations of the dependent geometric distribution and the geometric distribution use the deal and the roll functions respectively.

When ten successive cards are drawn from a shuffled bridge deck, the probabilities for obtaining the first ace are:

```
       4 52 DGD 10
 1 0.0769
 2 0.0724
 3 0.0681
 4 0.0639
 5 0.0599
 6 0.0561
 7 0.0524
 8 0.0489
 9 0.0456
10 0.0424
```

Since the first ace must appear by the forty-ninth card, the probabilities are mutually exclusive and exhaustive when ω of *DGD* is 49.

There are four 0s and forty-eight 1s in the result of the comparison 4<⍳52. A 0 represents a success, as it does in the geometric distribution. Multiple uses of the deal function simulate 25 samples of drawing the first ace from a shuffled bridge deck:

120/ The First Ace

```
FT 1++/∧\4≤25 52ρ±199ρ'(52?52),'
 2  2
 3  1
 4  3
 6  4
 7  2
 9  1
10  1
13  1
15  1
21  1
22  1
24  2
25  1
27  1
28  1
29  1
41  1
```

The first ace appeared as the fourth card in 3 of the 25 trials. In the 25 trials, the first ace never appeared as the first, fifth, eighth (and so on) card.

EXERCISES

15.1 In a diner, boxes containing 100 packages of jam have random assortments of 20 strawberry, 20 grape, 20 orange marmalade, 20 raspberry and 20 boysenberry. If 6 packages are randomly put in a bowl for each table, what is the probability distribution of the number of packages of raspberry likely to be served at a table?

15.2 Simulate 50 trials of the situation described in 15.1.

15.3 Over a long period of time, it is found that 20 percent of the customers in a diner order only a cup of coffee. What is the theoretical probability distribution for the number of people out of the next six customers to order only coffee?

15.4 Simulate 50 trials of the situation described in 15.4.

15.5 Suppose a new box of the 100 packages of jam described in 15.1 is opened and each customer at the counter is served 1 package with breakfast. What is the probability distribution for each of the first 6 customers getting the first package of raspberry?

15.6 Simulate 50 trials of the situation described in 15.5.

15.7 At the diner described in 15.3, what is the probability for each of the next 6 customers to be the first to order only a cup of coffee?

15.8 Simulate 50 trials of the situation described in 15.7.

16/ THE BIRTHDAY PROBLEM

$MD:TA1\ 1-x\backslash1-(\iota\alpha)\div\omega$
$MATCH:MP?\alpha\rho\omega$
$MP:1++/\wedge\backslash(\omega\iota\omega)=\iota\rho\square\leftarrow\omega$

What is the probability that two or more people in a room were born on the same day of the week? One way to approach this problem is to consider the probability that two or more individuals' birthdays did *not* fall on the same weekday.

There is a probability of 7÷7 or 1 that the first person considered was born some day in the week. The probability that the second person does not have the same birthday as the first is:

    x/7 6÷7
0.8571

The probability that the first three people have different birthdays is:

    x/7 6 5÷7
0.6122

The probability distribution for no matches among the first 1, 2, 3, 4, 5, 6 or 7 people consists of the corresponding elements in the vector result:

    x\7 6 5 4 3 2 1÷7
1 0.8571 0.6122 0.3499 0.1499 0.04284 0.00612

It is easier to generalize from the following expression:

    x\1-(ι7)÷7
1 0.8571 0.6122 0.3499 0.1499 0.04284 0.00612

Subtracting each of the probabilities that no two birthdays are alike from 1 gives the complementary probability distribution. Thus, the chances that at least 2 out of 7 people were born on the same day of the week are:

```
    1-x\1-(ι7)÷7
0  0.1429  0.3878  0.6501  0.8501  0.9572  0.9939
```

The results are not mutually exclusive as they were in all of our earlier distributions. The likelihood of a match increases as the number of people increases. A match must occur when all the unique possibilities have been exhausted. This is called a maturing or cumulative distribution. We can define a *matching distribution* function α*MD*ω. α is the total number of elements we are considering. ω is the number of unique possibilities for each element:

```
MD:TA1  1-x\1-(ια)÷ω
```

We can generate a table of the successive probabilities that at least 2 out of 10 people's birthdays occurred on the same weekday:

```
    10 MD 7
 1  0.0000
 2  0.1429
 3  0.3878
 4  0.6501
 5  0.8501
 6  0.9572
 7  0.9939
 8  1.0000
 9  1.0000
10  1.0000
```

A match must occur when 8 people are considered because there are only 7 different days in a week. Note that the number of different possibilities for each person remains constant, and that the events are independent of one another.

Let's look at an extension of the birthday problem. Given a roomful of people, what is the probability that at least 2 were born on the same day of the year? We take 365 unique days on which a birthday might occur (disregard the 29th of February) and use the matching distribution function to generate a distribution of the probabilities:

```
      10 MD 365
 1  0.0000
 2  0.0027
 3  0.0082
 4  0.0164
 5  0.0271
 6  0.0405
 7  0.0562
 8  0.0743
 9  0.0946
10  0.1169
```

The likelihood of a match is 0.0164 when there are 4 people in a room and 0.1169 when there are 10 present.

Suppose we generate a specific sequence of "birthdays" for five people and compare them for duplicates using the defined function $\alpha MATCH \omega$. The number in ω is the number of different possibilities for each element and α is the number of selected elements. With respect to the birthday problem, ω is the number of days in the year and α is the number of people in the sample.

$MATCH:MP?\alpha\rho\omega$

The expression $?\alpha\rho\omega$ generates a vector of birthdays:

```
      ?5ρ365
223 82 349 82 219
```

This vector is then used three times in the *match procedure* function $\alpha MP \omega$.

$MP:1++/\wedge\backslash(\omega\iota\omega)=\iota\rho\square\leftarrow\omega$

When $\square$ is the left argument of specification, the computer displays the array in ω. This allows us to see the randomly generated vector that is being analyzed when $\alpha MATCH \omega$ is executed.

Continuing with the same randomly created vector of birthdays, MP generates a counting vector based on the number of elements in ω:

```
     ιρ223 82 349 82 219
0 1 2 3 4
```

The expression ω⍳ω produces a vector containing the least index of each element of ω in ω:

```
      223 82 349 82 219⍳223 82 349 82 219
0 1 2 1 4
```

Then, the least index vector is compared to the counting vector using the equals function:

```
      0 1 2 1 4=0 1 2 3 4
1 1 1 0 1
```

The first 0 in the result represents the first match. That is, the fourth randomly generated birthday is the first duplicate in this particular simulation. Finally, *MP* calculates which randomly generated element represents the first matching birthday:

```
      1++/∧\1 1 1 0 1
4
```

Here's how this simulation would display at your terminal:

```
      5 MATCH 365
223 82 349 82 219
4
```

The first row of numbers is the randomly produced birthdays. The 4 in the second line indicates that the fourth one is the first to match another. Consider the following simulation:

```
      30 MATCH 365
214 36 138 108 161 160 203 43 78 94 96 168 44
    302 80 45 166 269 144 170 54 61 99 149 75
    256 364 348 242 202
31
```

None of the 30 randomly generated birthdays in this particular simulation match. Thus, the comparison using the equals function generates only 1s and the result of *MATCH* is 31, one *more* than the number specified in its left argument.

You can use the match function to generate repeated simulations and compare them with the theoretical results produced by the matching distribution function.

```
      30 MATCH 365
32 171 149 15 81 61 168 43 65 301 221 292 223
       117 35 185 135 323 18 198 171 131 353
       29 351 305 333 119 252 201
21

      30 MATCH 365
184 363 259 201 281 341 231 260 260 42 170 311
       171 217 47 102 123 201 242 251 143 260 47
       289 339 269 160 112 120 197
9
```

If you've studied the material in these 16 chapters, the probability should be close to 1 that you now have enough of a command of APL to find the theoretical probability of success or failure for events that interest you. Simulating those situations will give you confidence in the theoretical predictions. Enjoy experimenting!

EXERCISES

16.1 Make a guess at how many people would have to be in the same room in order that the probability that two or more would have the same birthday is as close to 0.5 as possible.

16.2 Experiment with the defined function to find the number exactly.

16.3 Ask 30 people when their birthdays are and record when, if at all, you obtain a pair.

16.4 Make a guess at how many people would be necessary to increase your certainty to 70 percent and then find that number exactly.

16.5 A function developed by Howard Peelle serves as an exercise in reading APL. It produces a visually pleasing result. Enter the functions $PO: \omega \top \iota \times / \omega$ and $TR: \omega \wedge \omega \neq ^{-}1 \Phi \omega$ and then execute the expression ' ○'$[TR\ PO\ 4\rho 2]$.

The Birthday Problem /127

16.6 The final challenge is to deal the shuffled deck you obtained in exercise 13.9 to produce a 3-dimensional array. Here's an example of a possible random result:

*T*97*Q*4*J*82*J*2652
HCSCHSCCDHCSD

5*A*4*KK*5*T*57378*A*
HDSCDDCCHDCDS

*AKQ*9*Q*36*K*4636*J*
HHSDHSDSDSCHC

7948*J*29*QATT*38
DSCHHSHDCSDHS

APPENDIX A: DEFINING FUNCTIONS

Using the program diskette

Before you do anything else, make a copy of the diskette, and put the original aside as a backup copy. Do your work with the copy you made.

The two workspaces on the diskette are called *PROB* and *DDEF*. The workspace *PROB* contains all of the defined functions shown in the text as well as the functions supporting the direct definition of functions. The workspace *DDEF* contains only the functions supporting direct definition.

To gain practice in defining functions, load the workspace called *DDEF* by making one of the following entries according to the requirements of your APL system:

)*LOAD* 0 *DDEF*

or

)*LOAD* 1 *DDEF*

The system will respond by displaying the date and time when this workspace was last saved. The *DEF* function in this workspace allows you to enter your function in direct definition form. Simply type *DEF* and press the key marked RETURN, ENTER or ⏎ on your keyboard. The system will wait for your next entry. Type the name of your function followed by a colon and the APL expression which defines it. Press the RETURN, ENTER or ⏎ key to transmit this entry to the computer. For example:

    DEF
$PO: \omega\top\iota\times/\omega$

You can enter each of the functions in the text in exactly the form in which it is shown; that is, in direct definition representation. You can name a copy of this workspace containing the new functions you've defined and save it by using the facilities provided by your particular APL system.

Using the del or canonical form of definition

You can always define a function on an APL computer by entering the *del* ∇ character followed by the function's header containing its name, argument(s) and result. After you press the RETURN, ENTER or ↵ key, the computer responds by displaying a line number in brackets and waits for you to enter the first line of your function's definition:

```
        ∇RESULT←NAME ARGUMENT
[1]
```

It is possible to write functions containing several lines. Thus, the computer's response after you've entered the first line is to display [2] and wait for you to enter the second line of your definition. To signal the computer that you want to leave definition mode, end a line with the del character, or simply enter a ∇ alone when the system prompts you with a new line number. This mode of definition is called the *del* or *canonical* form of function definition.

All of the functions in this text are expressed in the *direct* definition representation. If your system does not support direct definition, you can use the following models and make the substitutions described in order to define your functions in the canonical form; or, you can create your own direct definition workspace by entering the functions and variables listed in the next section.

The model for the canonical representation of a monadic function is:

```
        ∇Z←NAME W
[1]     Z←EXPRESSION∇
```

Using this model as a guide, the canonical representation of the possible outcomes function

$PO: \omega \top \iota \times / \omega$

is:

```
        ∇Z←PO W
[1]     Z←W⊤ι×/W∇
```

That is, you substitute *PO* for *NAME* and *W* for ω. Note that using *Z* and *W* to represent the result and the right argument was an arbitrary choice; any names could be used. The APL expression to the right of the colon in the direct definition form replaces the word *EXPRESSION* in the model. Note that the canonical form of representation begins and ends with the ∇ character.

130/ **Appendix A: Defining Functions**

Similarly, the dyadic function

$POC:\alpha[(PO\omega)+CM\omega]$

can be represented canonically by substituting A for α and W for ω in the following model:

```
       ∇Z←A NAME W
[1]    Z←EXPRESSION∇
```

Thus, the del representation of POC is:

```
       ∇Z←A POC W
[1]    Z←A[(PO W)+CM W]∇
```

As you complete portions of your work, you can assign your active workspace a name and save a copy of it according to the conventions of your APL system. At some future time you can use your defined functions or define additional ones by loading a copy of the saved workspace. The sequence of loading, doing additional work and then saving the workspace is used over and over as you begin to utilize the full capacity of the computer.

You should become familiar with the manuals that accompany the particular system you are using so that you become aware of further modifications which might be necessary. On any system, defining functions is the process by which you extend the APL language to get specific results appropriate in your area of work.

Creating your own direct definition workspace

If your APL system does not support direct definition, you can enter the following functions and variables to build your own *DDEF* workspace:

```
       ∇D←F9 E;F;I;K
[1]    F←(,(E='ω')∘.≠5↑1)/,E,(⌽4,ρE)ρ' Y9 '
[2]    F←(,(F='α')∘.≠5↑1)/,F,(⌽4,ρF)ρ' X9 '
[3]    F←1+ρD←(0,+/¯6,I)↓(-(3×I)++\I←':'=F)⌽F,(⌽6,ρF)ρ' '
[4]    D←3⌽C9[1+(1+'α'∊E),I,0;],⍋D[;1,(I←2↓⍳F),2]
[5]    K←K+2×K<1⌽K←I∧K∊(>⌿ 1 0 ⌽'←⎕'∘.=E)/K←+\~I←E∊A9
[6]    F←(0,1+ρE)⌈ρD←D,(F,ρE)↑⍟ 0 ¯2 ↓K⌽' ',E,[1.5] ';'
[7]    D←(F↑D),[1] F[2]↑'A',E
[8]    ∇
```

```
        ∇DEF;⎕IO
[1]     ⎕IO←1
[2]     0ρ⎕FX F9 ⎕
[3]     ∇

A9←'0123456789ABCDEFGHIJKLMNOPQRSTUVWXYZ⎕∆'

C9←5 11ρ'⎕⎕⎕Z9←⎕⎕⎕⎕⎕⎕Y9Z9←⎕⎕⎕⎕⎕⎕Y9Z9←X9⎕⎕⎕)/3→(0=1↑,⎕⎕⎕⎕→0,0ρZ9←'
```

Note: You should enter a blank for each ⎕ shown in the value of the variable C9. The ⎕ characters are printed here only as an aid in counting the number of blanks required in the specification.

The direct definition compiler listed in this section is based on the one shown in "Notation as a Tool of Thought," Kenneth E. Iverson, *A Source Book in APL,* Adin D. Falkoff and Kenneth E. Iverson, APL PRESS, Palo Alto, 1981. Some modifications suggested by Roland Pesch have been incorporated in the value of the variable A9.

APPENDIX B: DEFINED FUNCTIONS

The following is an alphabetized list of the defined functions used in this booklet and the page number on which each is first defined:

all combinations, 50
  $AC:(((2\times\rho\omega)\rho 0\ 1)\backslash\omega)POC(\rho\omega)\rho 2$

binomial distribution of probabilities, 97
  $BD:TA0\alpha BP\omega$

binomial probabilities, 72
  $BP:((\iota 1+\alpha)!\alpha)\times(\omega*\iota 1+\alpha)\times(1-\omega)*\phi\iota 1+\alpha$

correction matrix, 40
  $CM:\lozenge(\phi\rho PO\omega)\rho^{-}1\downarrow+\backslash 0,\omega$

correction matrix (alternate), 46
  $CMALT:(^{-}1\downarrow+\backslash 0,\omega)\circ.+(\times/\omega)\rho 0$

combinations, 76
  $COMB:\lozenge((\alpha!\omega),\alpha)\rho(,\alpha CR\omega)/,(\rho\alpha CR\omega)\rho\iota\omega$

combinations as rows, 75
  $CR:\lozenge(\alpha=+\neq PO\omega\rho 2)/PO\omega\rho 2$

dependent binomial distribution, 104
  $DBD:TA0((\iota 1+\omega)!1\uparrow\alpha)\times((\phi\iota 1+\omega)!|-/\alpha)\div\omega!1\downarrow\alpha$

dependent geometric distribution, 118
  $DGD:TA1((^{-}1+1\uparrow\alpha)!\omega\uparrow\phi\iota 1\downarrow\alpha)\div!/\alpha$

frequency procedure, 60
  $FP:(ONUB\omega),[.5]+/(ONUB\omega)\circ.=\omega$

frequency table, 60
  $FT:FP,\omega$

frequency table not ordered, 72
  $FTN:FTNP,\omega$

frequency table not ordered procedure, 72
  FTNP:(NUBω),[.5]+/(NUBω)∘.=ω

geometric distribution of probabilities, 97
  GD:TA1αGPω

geometric probabilities, 96
  GP:α×(1-α)*ιω

histogram, 69
  HIST:' ⎕'[((ι1+ω)!ω)∘.>ι⌈/(ι1+ω)!ω]

matching birthdays simulation, 125
  MATCH:MP?αρω

matching distribution, 124
  MD:TA1 1-×\1-(ια)÷ω

match procedure, 125
  MP:1++/∧\(ωιω)=ιρ⎕←ω

nub procedure, 58
  NP:((ωιω)=ιρω)/ω

number of permutations, 85
  NPERM:(α!ω)×!α

unique elements, 58
  NUB:NP,ω

ordered nub, 60
  ONUB:(NUBω)[⍋NUBω]

permutations, 85
  PERM:(α=+/∨≠(POαρω)∘.=ιω)/POαρω

permutations (alternate), 92
  PERMALT:UO⍀POα↑ϕ1+ιω

possible outcomes, 34
  PO:ωτι×/ω

possible outcomes as characters, 40
 $POC:\alpha[(PO\omega)+CM\omega]$

Pascal triangle, 66
 $PT:(\iota 1+\omega)\circ.!\iota 1+\omega$

tables in origin 0, 97
 $TA0:0\ 0\ 0\ 4\top(\iota\rho\omega),[.5]\omega$

tables in origin 1, 97
 $TA1:0\ 0\ 0\ 4\top(1+\iota\rho\omega),[.5]\omega$

unique orderings, 88
 $UO:\omega+\alpha\leq\omega$

Chronological list of defined functions

Page	Function
34	$PO:\omega\top\iota\times/\omega$
40	$CM:\mbox{\textcircled{Q}}(\phi\rho PO\omega)\rho^{-}1\downarrow+\backslash 0,\omega$
	$POC:\alpha[(PO\omega)+CM\omega]$
46	$CMALT:(^{-}1\downarrow+\backslash 0,\omega)\circ.+(\times/\omega)\rho 0$
50	$AC:(((2\times\rho\omega)\rho 0\ 1)\backslash\omega)POC(\rho\omega)\rho 2$
58	$NUB:NP,\omega$
	$NP:((\omega\iota\omega)=\iota\rho\omega)/\omega$
60	$ONUB:(NUB\omega)[\triangle NUB\omega]$
	$FT:FP,\omega$
	$FP:(ONUB\omega),[.5]+/(ONUB\omega)\circ.=\omega$
66	$PT:(\iota 1+\omega)\circ.!\iota 1+\omega$
69	$HIST:'\ \square'[((\iota 1+\omega)!\omega)\circ.>\iota\lceil/(\iota 1+\omega)!\omega]$
72	$FTN:FTNP,\omega$
	$FTNP:(NUB\omega),[.5]+/(NUB\omega)\circ.=\omega$
	$BP:((\iota 1+\alpha)!\alpha)\times(\omega\star\iota 1+\alpha)\times(1-\omega)\star\phi\iota 1+\alpha$
75	$CR:\mbox{\textcircled{Q}}(\alpha=+\neq PO\omega\rho 2)/PO\omega\rho 2$
76	$COMB:\mbox{\textcircled{Q}}((\alpha!\omega),\alpha)\rho(,\alpha CR\omega)/,(\rho\alpha CR\omega)\rho\iota\omega$
85	$PERM:(\alpha=+/\vee\neq(PO\alpha\rho\omega)\circ.=\iota\omega)/PO\alpha\rho\omega$
	$NPERM:(\alpha!\omega)\times!\alpha$
88	$UO:\omega+\alpha\leq\omega$
92	$PERMALT:UO\backslash PO\alpha\uparrow\phi 1+\iota\omega$
96	$GP:\alpha\times(1-\alpha)\star\iota\omega$
97	$TA0:0\ 0\ 0\ 4\mbox{\textcircled{$\top$}}(\iota\rho\omega),[.5]\omega$
	$TA1:0\ 0\ 0\ 4\mbox{\textcircled{$\top$}}(1+\iota\rho\omega),[.5]\omega$
	$BD:TA0\alpha BP\omega$
	$GD:TA1\alpha GP\omega$
104	$DBD:TA0((\iota 1+\omega)!1\uparrow\alpha)\times((\phi\iota 1+\omega)!|-/\alpha)\div\omega!1\downarrow\alpha$
118	$DGD:TA1((^{-}1+1\uparrow\alpha)!\omega\uparrow\phi\iota 1\downarrow\alpha)\div!/\alpha$
124	$MD:TA1\ 1-\times\backslash 1-(\iota\alpha)\div\omega$
125	$MATCH:MP?\alpha\rho\omega$
	$MP:1++/\wedge\backslash(\omega\iota\omega)=\iota\rho\square\leftarrow\omega$

Appendix B: Defined Functions

APPENDIX C: ANSWERS

2.1 ?5
2.2 ?8
2.3 ?8 8 8 8 8 8 8 8 8 8 8
2.4 ?4 4 4 4 4
2.5 ?4 4 8 8
2.6 ?4 8 4 8
2.7 ?5
2.8 ?4 4 4 4 4 4 4 4 4 4

3.1 ?4
3.2 ?5ρ4
3.3 ?100ρ4
3.4 ?2 10ρ4   or   ?10 2ρ4
3.5 ?5 3 6ρ4
3.6 ?6 2ρ3 4
3.7 ?10 2ρ4 5

4.1 'TH'[?4 16ρ2]
4.2 'MTWRF'[?5]
4.3 'A23456789TJQK'[?5ρ13]
4.4 'COB'[?3 27ρ3]
4.5 TRUMPET
4.6 'ROYGBVI'[?4ρ7]
4.7 'PNDQ'[?5ρ4]
4.8 'ABCDEFGHIJKLMNOPQRSTUVWXYZ'[?25 5ρ26]
4.9 'O▲□↑↓ΦΨ|'[?20 3ρ8]
4.10 'O▲□↑↓ΦΨ|'[?2 10 3ρ8]

5.1 PO 2 2 2 2 2
5.2 'TH'[PO 2 2 2 2 2]
5.3 'TH'[?5 32ρ2]
5.4 PO 5 2 3
5.5 ×/10ρ4
5.6 ×/4ρ26
5.7 'YRWP'[PO 4 4 4]

6.1 'GYRWB' POC 2 3
6.2 'RWBGY' POC 3 2
6.3 'TH123456' POC 2 6
6.4 'RGLSRRGLSR' POC 2 3 2 3

6.5 `'A23456789TJQKCDHS' POC 13 4`

7.1 `AC 'ABCDE'`
7.2 `2*5`
7.3 `AC 'ABCDEF'`
7.4 `2*6`
7.5 `AC 'ESAM'`
7.6 `2*4`

8.1 `(+/(ONUB+/PO 2 2 2)∘.=+/PO 2 2 2)÷2*3`
8.2 `FT+/?3 800ρ2`
8.3 `100×1 3 3 1`
8.4 `(+/(ONUB+/PO 2 2 2 2 2)∘.=+/PO 2 2 2 2 2)÷2*5`
8.5 `FT+/?5 3200ρ2`
8.6 `100×1 5 10 10 5 1`
8.7 `CMALT2:(×/ω)/(φ1,ρω)ρ⁻1↓+\0,ω`

9.1 `(2!4)÷2*4`
9.2 `(3!5)÷2*5`
9.3 `AC 'SHARED'`
9.4 `(ι7),[.5]((ι7)!6)÷2*6`
9.5 `HIST 6`
9.6 `(4!6)÷2*6`
9.7 `5 BP .6`
9.8 `FT+/6>?5 100ρ10`

10.1 `4!5`
10.2 `'PSEJT'[4 COMB 5]`
10.3 `'PSEJT'[1 COMB 5]`
10.4 `'PSEJT'[5 COMB 5]`
10.5 `'CHFLMP'[4 COMB 6]`
10.6 `(4!6)÷2*6`
10.7 `'MNOPQRS'[4 COMB 7]`

11.1 `⌽'CORN'[4 PERM 4]` and `⌽'ACORN'[5 PERM 5]`
11.2 `3 NPERM 5`
11.3 `'ACORN'[3 PERM 5]`
11.4 `'RBYG'[4 PERM 4]`
11.5 `4 NPERM 6`
11.6 `5 NPERM 52`
11.7 `3 NPERM 5`
11.8 `+/1 2 3 4 5 NPERM 5`
11.9 `×/9 8 7 6` or `3024` for `PERM`

compared with 4!9 or 126 for *PERMALT*

12.1 4 *BD* .05
12.2 *FT*+⌿5>?4 50ρ100
12.3 .05 *GD* 10
12.4 *FT* 1++⌿∧⍀5≤?10 200ρ100

13.1 20 50 *DBD* 10
13.2 *FT* 1++/∧\20≤40 10ρ⍟319ρ'(10?50),'
13.3 10 *BD* .4
13.4 *FT*+⌿2>?10 40ρ5
13.5 4 *BD* 1÷13
13.6 *FT*+⌿1>?4 100ρ13
13.7 4 52 *DBD* 13
13.8 *FT*+/4>100 13ρ⍟799ρ'(13?52),'
13.9 ('A23456789TJQKCDHS' *POC* 13 4)[;52?52]

14.1 6!11
14.2 (!12)÷×/!5 3 4
14.3 8!18 or 10!18
14.4 (4!4 5 6 7 8)÷2×+/4!4 5 6 7 8
14.5 *FT* 1++⌿5>+⍀?9 60ρ2

15.1 20 100 *DBD* 6
15.2 *FT*+/20>50 6ρ⍟399ρ'(6?100),'
15.3 6 *BD* .2
15.4 *FT*+⌿1>?6 50ρ5
15.5 20 100 *DGD* 6
15.6 *FT* 1++/∧\20≤50 6ρ⍟399ρ'(6?100),'
15.7 .2 *GD* 6
15.8 *FT* 1++⌿∧⍀1≤?6 50ρ5

16.2 23 *MD* 365
16.4 30 *MD* 365
16.6 0 2 1⌽4 13 2ρ⌽('A23456789TJQKCDHS' *POC* 13 4)[;52?52]

Appendix C: Answers /139

APPENDIX D: APL INDEX

absolute value, $|\omega$, 105
and scan, $\wedge\backslash\omega$, 100
assignment, see specification
axis operator, $f[\omega]$, 61

binomial coefficients, $\alpha!\omega$, 66
binomial combinations outer product, $\alpha\circ.!\omega$, 66

canonical form of function definition, 130, 131
catenate, α,ω, 41, 61
character representation, $\alpha\top\omega$, 96
)$CLEAR$ system command, 12
colon, : , 35

deal, $\alpha?\omega$, 106
del, ∇ , 130
del form of function definition, see canonical form
direct form of function definition, 129-131
divide, $\alpha\div\omega$, 12
drop, $\alpha\downarrow\omega$, 41

encode, $\alpha\top\omega$, 29, 32-33
equals, $\alpha=\omega$, 57
equals outer product, $\alpha\circ.=\omega$, 61
execute, $\pm\omega$, 107
expand, $\alpha\backslash\omega$, 51

factorial, $!\omega$, 81
floor, $\lfloor\omega$, 98

greater than outer product, $\alpha\circ.>\omega$, 69

index generator, $\iota\omega$, 33
index of, $\alpha\iota\omega$, 59
index origin, $\Box IO$, 25, 59, 61
indexing, $\alpha[\omega]$, 24, 60

laminate, $\alpha,[X]\omega$, 61
less than or equal to, $\alpha\le\omega$, 99
)$LOAD$ system command, 129

logical and, α∧ω, 99
logical or, α∨ω, 84

magnitude, see absolute value
maximum, α⌈ω, 68
maximum reduction, ⌈/ω, 68
minus reduction, -/ω, 105

not equals, α≠ω, 57

or reduction, ∨⌿ω, 84
outer product, ∘.g, 44

plus outer product, α∘.+ω, 44
plus reduction, +/ω, 34, 61
plus reduction along the first axis, +⌿ω, 56
power, α*ω, 53
print precision, ⎕PP, 67
product operator, f.g, 44

quote, ' , 23

ravel, ,ω, 58
reciprocal, ÷ω, 14
reduction along the first axis, f⌿ω, 56
reduction operator, f/ω, 34, 87-92
replicate, α/ω, 59
reshape, αρω, 19-22
reverse, ⌽ω, 42
roll, ?ω, 15-17

scan operator, f\ω, 41, 87-92
shape, ρω, 42
specification, α←ω, 25, 125

take, α↑ω, 92
times reduction, ×/ω, 34
transpose, ⍉ω, 43

upgrade, ⍋ω, 60